Palgrave Studies in Impact Finance

Palgrave Studies in Green Finance

Series Editors
Helen Chiappini, Dipartimento di Economia Aziendale, D'Annunzio
University of Chieti–Pescar, PESCARA, Pescara, Italy
Mario La Torre, Facolta di Economia, Universita La Sapienza, ROMA,
Roma, Italy

This subseries explores studies on green finance, specifically with regards to green investments, business models, roles of different actors in the green finance market, new regulatory and disclosure trends, green assets risks and performance alongside emerging areas including climate risk and green fintech with both theoretical and empirical approaches. Palgrave Studies in Green Finance represents the first international series dedicated to such a topic, meeting the growing interest of scholars, policy makers, regulators, and students in green finance.

Alessandro Rizzello

Green Investing

Changing Paradigms and Future Directions

Alessandro Rizzello
Department of Law, Economics and Sociology
University "Magna Græcia" of Catanzaro
Catanzaro, Italy

ISSN 2662-5105　　　　　　ISSN 2662-5113　(electronic)
Palgrave Studies in Impact Finance
ISSN 2662-7388　　　　　　ISSN 2662-7396　(electronic)
Palgrave Studies in Green Finance
ISBN 978-3-031-08030-2　　ISBN 978-3-031-08031-9　(eBook)
https://doi.org/10.1007/978-3-031-08031-9

© The Author(s), under exclusive license to Springer Nature Switzerland AG 2022

This work is subject to copyright. All rights are solely and exclusively licensed by the Publisher, whether the whole or part of the material is concerned, specifically the rights of translation, reprinting, reuse of illustrations, recitation, broadcasting, reproduction on microfilms or in any other physical way, and transmission or information storage and retrieval, electronic adaptation, computer software, or by similar or dissimilar methodology now known or hereafter developed.

The use of general descriptive names, registered names, trademarks, service marks, etc. in this publication does not imply, even in the absence of a specific statement, that such names are exempt from the relevant protective laws and regulations and therefore free for general use.

The publisher, the authors, and the editors are safe to assume that the advice and information in this book are believed to be true and accurate at the date of publication. Neither the publisher nor the authors or the editors give a warranty, expressed or implied, with respect to the material contained herein or for any errors or omissions that may have been made. The publisher remains neutral with regard to jurisdictional claims in published maps and institutional affiliations.

Cover image: © Andriy Onufriyenko/Getty images

This Palgrave Macmillan imprint is published by the registered company Springer Nature Switzerland AG
The registered company address is: Gewerbestrasse 11, 6330 Cham, Switzerland

*to my father and my mother with whom
I would have liked so much to share
the joy for the publication of this book*

Foreword

Green Finance (GF) is heralded in theory and practice as the new panacea; as the ideal way to support the green transition of businesses into more sustainable, environmentally responsible forms by means of incentivized financial investments. It is thus no wonder that regulators around the globe, a recent example being the European Union, are proposing new legislation to support this concentrated effort. Recent initiatives include the Green New Deal (GND), the EU Sustainable Finance Action Plan, and more specifically the EU Green Bond Standard (GBS). Similar developments can be witnessed in many countries and areas, particularly in North America, China, and Oceania. Following this momentum, an increasing number of regional public bodies, financial service providers such as banks or specialized investment firms offer innovative green finance instruments and investment portfolios to support the intended green transition on the ground.

The basis of Green Finance comprises green bonds and loans as part of impact investments that are either directly linked to projects or ventures that aim for a positive environmental impact or are used more generally to improve the overall ESG (Environmental, Social, and Governance) scores of the fund-seeking companies through changes in business operations. In the latter case, the interest rate is often linked to specific outcomes and is even reduced when certain Environmental, Social, and Governance (ESG) score thresholds are reached. However, besides these general incentivization principles, there is no comprehensive theory nor standardization of

GF, and recent studies on financial covenants of green bonds bring to light little resemblances to the promises and green marketing activities of the issuers.

In fact, to date, we know little about the actual motivations to issue or buy GF instruments, their impact on other instruments, the pricing mechanisms, the risk/reward profiles, the setup of suitable financial covenants, or even on the ideal metrics to measure the promised environmental impacts of the projects. As a result, the lack of a comprehensive theory raises serious questions about the credibility and effectiveness of the individual GF instruments—and thus the overall market—and leads to accusations of greenwashing or lackluster engagement strategies.

Following recent announcements of green investment opportunities by major funds, banks and investment companies that all carry the emblem of green finance, the question emerges whether the deliberate ambiguity of the whatness of green investments in the market may not simply disguise old wine in new bottles. What may be worse is that this might take away funds from smaller but highly dedicated impact investing initiatives by less powerful actors, as now virtually all gets labeled as green, with green investments becoming, what Levi-Strauss and Roland Barthes seem to regard as "floating" or "empty" signifiers, despite having quite powerful—albeit highly questionable—impacts for sign-users. This leads to a dilution of green capital and may ultimately reduce the desired impact of such green capital allocation.

Engaging with these growing tensions, this book provides much-needed, excellent guidance on the various instruments for green finance and their connection to the Sustainable Development Goals (SDGs). What is more, it does not shy away from looking deeper at the serious issue of greenwashing and presents potential remedies. Finally, it looks how green investments, if guided carefully, can become the foundation of a strong performativity toward a greener development of the economy after the Covid crises.

I wish the reader a most pleasant journey and many insights from reading this book.

Prof. Dr. Othmar M. Lehner
Hanken School of Economics
Helsinki, Finland

Acknowledgments

First and foremost, I would like to express my gratitude to Prof. Mario La Torre, Editor-in-Chief of the Palgrave Macmillan Studies in Impact Finance and to Prof. Helen Chiappini, sub-series editor of Palgrave Studies in Green Finance and to the anonymous reviewers for approving my book proposal and for their useful suggestions.

My gratitude extends to Prof. Annarita Trotta (Magna Graecia University of Catanzaro) for being my mentor over these years of research on sustainable finance and for constantly encouraging me to write this book.

Special thanks to many people who made significant contributions along my academic career with useful discussions on sustainable finance evolutions and on industry needs as well as through their support in co-authorship of my publications with valuable perspectives.

Moreover, I would like to show my gratitude to the most important people in my life. Thank you to my wife, Maria. Thank you for always believing in me and for having encouraged me and, overall, for your understanding and patience. I would also extend my thanks to all my brothers and all my little nephews, they have been my great supporters and special friends.

Finally, I would also express my gratitude to the Palgrave Macmillan team, and in particular to Ellie Duncan, Tula Weis, and Redhu Ruthroyoni for their support during the publication process.

Contents

1	Introduction	1
2	What's in a Name? Mapping the Galaxy of Green Finance	9
3	The Green Financing Framework Combining Innovation and Resilience: A Growing Toolbox of Green Finance Instruments	55
4	Green Finance and SDGs: Emerging Trends in the Design of Green Investment Portfolios	85
5	Beyond Greenwashing: An Overview of Possible Remedies	107
6	Financing the Green Recovery: The New Directions of Finance After the COVID-19 Crisis	133
7	Conclusions	159
Index		165

List of Figures

Fig. 1.1	Book research outline (*Source* Author's elaboration)	4
Fig. 2.1	Parallel evolution of green finance policies, green finance research and green finance definitions' focus (*Source* Author's elaboration)	15
Fig. 2.2	Green finance and other forms of finance for sustainability (*Source* Author's elaboration)	19
Fig. 2.3	Green finance definition's galaxy: a four-element visualization (*Source* Author's elaboration)	21
Fig. 2.4	The green finance ecosystem (*Source* Author's elaboration)	22
Fig. 2.5	A visualization of the green finance sectors through the four elements of nature (*Source* Author's elaboration)	41
Fig. 3.1	The green financing framework (*Source* Author's elaboration)	59
Fig. 3.2	The Environmental Impact Bond model (*Source* Author's elaboration)	70
Fig. 3.3	The Sustainability-Linked Bond model (*Source* Author's elaboration)	71
Fig. 4.1	A segmentation of SDGs by broader areas (*Source* Author's elaboration)	87
Fig. 4.2	The spectrum of green investors (*Source* Author's elaboration)	91
Fig. 5.1	Greenwashing in green finance (*Source* Author's elaboration)	122

Fig. 6.1	Determinants for the new role of finance after COVID-19 (*Source* Author's elaboration)	148
Fig. 7.1	A sysntesis of the book's research contribution (*Source* Author's elaboration)	161

List of Tables

Table 2.1	The green finance standard setters: an overview	32
Table 2.2	Green and sustainable bonds: an overview	37
Table 2.3	A segmentation of green loans	38
Table 2.4	Main investment approaches in green equity markets	40
Table 3.1	The use of proceeds in green bonds: the eligible green activities	61
Table 3.2	Main index providers in green finance, approach and some examples	63
Table 3.3	List of the leading crowdfunding platforms for green investments	66
Table 4.1	Relevant SDG Targets related to the Environment	88
Table 4.2	Major ESG rating and index firms per country	95
Table 4.3	The SDGs in green equity markets: some examples	99
Table 5.1	The use of green taxonomies in the green finance sector	117
Table 5.2	Illustrative cases of greenwashing in green finance	124

CHAPTER 1

Introduction

1.1 Green Investing: Setting the Scene

The financial industry is undergoing a fundamental transformation, following the prevailing global trend of turning to a more sustainable and environmentally responsible way of living and doing business (Weber, 2016). The need for action to reconcile economic development with environmental protection has highlighted green finance as a commonly used strategy for dealing with environmental problems (Wang & Zhi, 2016). In particular, the growing attention to green investments can be seen as a direct consequence of a series of milestones in the following fields: (*i*) policy (i.e., the Paris Agreement on Climate Change, the UN Sustainable Development Goals of the Agenda 2030, the launch of the European Union Green Deal Investment Plan); (*ii*) regulatory (the new European Union Taxonomy Regulation, the EU's new Non-Financial Reporting Directive, the Green Finance Strategy adopted in 2019 by the UK Government after Brexit); and (*iii*) socio-economic (the Fridays for Future movement, or the spread of the COVID-19 virus).

The convergence of such drivers has increased the adoption of environmentally friendly approaches to financing green business models and projects (Khalil & Nimmanunta, 2022). These transformative behaviors further signal a radical change of direction from how the financial industry has approached sustainability in the past. This transformation implies that the environmental investors' commitment may represent a viable way

© The Author(s), under exclusive license to Springer Nature Switzerland AG 2022
A. Rizzello, *Green Investing*, Palgrave Studies in Impact Finance, https://doi.org/10.1007/978-3-031-08031-9_1

to create long-term value for the investment itself, while also adding value for society by promoting environmental benefits (de la Orden & de Calonje, 2022; La Torre & Chiappini, 2020; Quatrini, 2021). The nexus between the issue of environmental protection and investment took place in the 1990s, when investors had increasingly begun to incorporate environmental factors into their investment decisions, by adopting screening criteria to identify their "environmentally responsible" investees (Chatzitheodorou et al., 2019). Furthermore, the incorporation of environmental risks within investment decisions (Epstein & Roy, 2017) aligned with this tendency; this evolved from a simple consideration of how environmental risks influence the counterparty's risk, and began to take account of financial systemic risks (Campiglio et al., 2018; Semieniuk et al., 2021). Therefore, financial actors now contribute to environmental sustainability both directly, through their targeted (use of proceeds) investments; and indirectly, through general/untargeted green investments (Migliorelli & Dessertine, 2019).

As a result, the green finance sector plays an increasingly important role in directing capitals to investments that help to more efficiently address environmental outcomes. At the same time, the growth of this sector is threatened by various challenges, such as regarding the management and measuring of environmental risks in the investment process, or the reputational issues due to the "greenwashing" risk, which could even compromise the legitimacy of the entire sector (Caldecott, 2020; Gilchrist et al., 2021). In this sense, several key changes are occurring in the regulatory actions at international level, which aim to overcome the limitations and fragmentation of voluntary non-financial sustainability disclosure frameworks and standards (Dikau & Volz, 2021; Fisch, 2018; Park, 2018). Finally, the global stimulus packages in response to the negative economic impact of the COVID-19 pandemic produced new policy-related and economic enabling conditions; this has accelerated a transformative scenario, confirming that investors are changing how they operate (Ionescu, 2021). In simple terms, both external and internal pressures are transforming the approach of investors, while they are also benefiting from the new business opportunities offered by the environmental protection principle.

1.2 The Research Aims

On a theoretical level, the theme of green finance is an emerging and rapidly growing field of research, concerned with the financial implications of the environmental protection measures for industries and firms, and the need to transition to a sustainable economy. However, the green finance academic landscape presents a higher level of fragmentation (Zhang et al., 2019) due to the lack of definitional clarity, as well as the abundance of tools, sub-sectors, and segments, which makes understanding and research exploration challenging (Andreeva et al., 2018; Berensmann & Lindenberg, 2016). At the same time, a holistic approach is needed to track and map the current (r)evolutions, in both theoretical and practical aspects; especially after the recent transformative regulatory and socio-economic events.

With these objectives in mind, the book intends to provide: (i) a map of the green finance sector, by highlighting its conceptual and market landscape; (ii) an updated understanding of the evolving landscape of green financial instruments; (iii) an overview of the emerging trends in the design of green investment portfolios, with a particular focus on SDG green themes; (iv) a conceptualization of the phenomenon of *greenwashing* and an analysis of possible solutions to mitigate such risk; and (v) an analysis of the role that green finance can play in the green COVID recovery, and of *green-led* paradigm shifts within the financial system.

1.3 Methodological Notes

This book focuses on the following topics:

- A contextualization of green finance within the diverse forms of finance offered for sustainability and related markets; outlining green finance features and components of the green financing framework and industry;
- A rigorous definition of green finance's theoretical boundaries;
- An exploration of the characteristics, investment strategy, and classification of green financing tools;
- A discussion of the emergent trends in green financing and an analysis of case studies;

- A synthetic assessment of the evolution in the design of green investing portfolios, conducted by analyzing the limits and potentials of innovative investment strategies;
- An exploration of determinants of the greenwashing phenomenon in green finance, and an identification of possible remedies;
- The tracking and analysis of transformative theoretical and practical drivers for new directions in finance, to support sustainable benefits for society and the environment.

This book is founded upon four major aspects that characterize current debates in green finance, in both theoretical and practical contexts: definitional issues, financial innovation, greenwashing, and divergences in the integration of environmental and non-financial returns.

In order to achieve these objectives, the book adopts a variety of methods, ranging from a systematic literature review to case studies, to provide the provision of theoretical overviews based also on triangulated analysis (such as in Chapters 2 and 6). The book supplements this analysis by building conceptual frameworks, through the assimilation and combination of evidence in the form of previously developed concepts and theories, and practical case studies. Thus, the book is organized to provide readers with clear theoretical assumptions and boundaries, as well as practical examples of green finance and investments, as depicted in the framework in Fig. 1.1.

In the following section, the individual book chapters are described in more detail, by outlining their core elements.

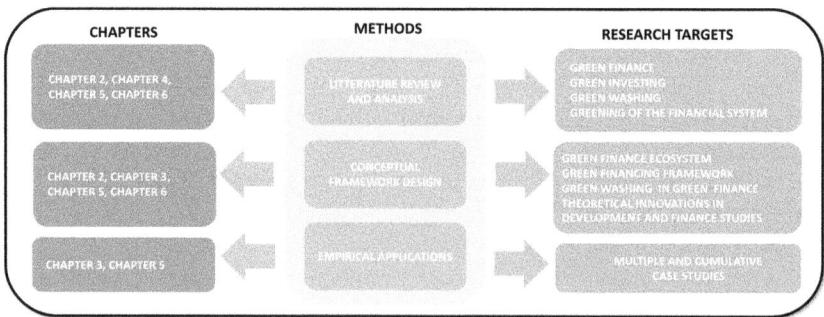

Fig. 1.1 Book research outline (*Source* Author's elaboration)

1.4 Outline of Chapters

The analysis is structured in five chapters, in addition to the introductory and concluding chapters. The second chapter introduces the topic of green finance. For this purpose, it summarizes the state of the art of this newly formed interdisciplinary field by setting out definitional boundaries and a conceptual framework, including the instruments, markets, and actors in this field. An analysis of the market ecosystem is also provided, based on the results of a literature review. Moreover, the chapter describes asset classes, including the green finance varieties and typologies of financial instruments employed, by highlighting the segmentation of the market, with a focus on recent regulation.

The third chapter examines how the landscape of green financial instruments is evolving in theory and in practice, and opens up useful directions for future research in this field. A case study approach is employed to demonstrate how alternative and previously unexplored green financial tools have been applied in environmentally oriented investment projects, by highlighting their similarities, differences, and potentials, as well as implications for the green finance market's development.

The fourth chapter explores the evolution in the design of green portfolios in the financial industry. By correlating emerging trends related to the fulfillment of green-themed Sustainable Development Goals, an overview is provided of the most important strategies adopted in designing green portfolios, as well as the role of the financial intermediation chain. The analysis highlights the issue of a green alignment strategy for portfolios.

The fifth chapter engages with the concept of greenwashing as an emerging issue in the financial world, with particular emphasis on the sustainable/green finance industry. Based on a literature review of the greenwashing topic, and a triangulated analysis that identifies the main features and origin of this phenomenon, two aims are pursued: first, to identify the main forms of greenwashing in green finance; and, second, to identify useful measures and tools that could be adopted to mitigate this risk.

The sixth chapter offers an integrated conceptual framework that brings together environmental sustainability and financial theory, and the *green-led* paradigm shifts in the financial system now occurring due to the pandemic. The results also refer to the previous chapters' theoretical and empirical findings, as a logical convergence of the book's analysis.

1.5 Research Innovation and Audience

The book provides a unique and up-to-date picture of green finance by highlighting, using both theoretical and practical lenses, the current changing paradigms and future directions in this field.

The analysis of findings and results presented in this text are expected to be useful for scholars and students in different academic disciplines (such as economics, finance, political science, and entrepreneurship) who are interested in green finance, and in the financing of green projects through more or less complex financial delivery tools. Moreover, leaders and staff of delivery organizations, intermediaries, foundations, as well as public bodies, may also find the book's analysis beneficial in addressing their specific needs.

In addition, the issues debated in this manuscript (such as the green alignment of portfolios, greenwashing, the adoption of global standards and frameworks, and the emerging green investment practices that incorporate financial innovations) also provide suggestions for further research in interconnected fields such as Sustainable Finance, Accounting, Entrepreneurial Finance, Public Finance, Financial Regulation and Policy, or Development Finance. In conclusion, the book will not only help researchers and policymakers to derive theoretical or practical implications; more broadly, it also captures the evolving complexity of the field, at the dawn of extraordinary, green-driven changes in policy programs and in the financial industry.

References

Andreeva, O. V., Vovchenko, N. G., Ivanova, O. B., & Kostoglodova, E. D. (2018). Green finance: Trends and financial regulation prospects. In *Contemporary issues in business and financial management in Eastern Europe*. Emerald Publishing Limited.

Berensmann, K., & Lindenberg, N. (2016). *Green finance: Actors, challenges and policy recommendations* (German Development Institute/Deutsches Institut für Entwicklungspolitik (DIE) Briefing Paper, 23).

Caldecott, B. (2020). Post Covid-19 stimulus and bailouts need to be compatible with the Paris Agreement. *Journal of Sustainable Finance & Investment*, 1–8.

Campiglio, E., Dafermos, Y., Monnin, P., Ryan-Collins, J., Schotten, G., & Tanaka, M. (2018). Climate change challenges for central banks and financial regulators. *Nature Climate Change*, 8(6), 462–468.

Chatzitheodorou, K., Skouloudis, A., Evangelinos, K., & Nikolaou, I. (2019). Exploring socially responsible investment perspectives: A literature mapping and an investor classification. *Sustainable Production and Consumption, 19*, 117–129.

de la Orden, R., & de Calonje, I. (2022). *Sustainability-linked finance: Mobilizing capital for sustainability in emerging markets* (EMCompass; Note 110). International Finance Corporation.

Dikau, S., & Volz, U. (2021). Central bank mandates, sustainability objectives and the promotion of green finance. *Ecological Economics, 184*, 107022.

Epstein, M. J., & Roy, M. J. (2017). Integrating environmental impacts into capital investment decisions. In *The green bottom line* (pp. 100–114). Routledge.

Fisch, J. E. (2018). Making sustainability disclosure sustainable. *The Georgetown Law Journal, 107*, 923.

Gilchrist, D., Yu, J., & Zhong, R. (2021). The limits of green finance: A survey of literature in the context of green bonds and green loans. *Sustainability, 13*(2), 1–12, 478.

Ionescu, L. (2021). Leveraging green finance for low-carbon energy, sustainable economic development, and climate change mitigation during the COVID-19 pandemic. *Review of Contemporary Philosophy, 20*, 175–186.

Khalil, M. A., & Nimmanunta, K. (2022). Conventional versus green investments: Advancing innovation for better financial and environmental prospects. *Journal of Sustainable Finance & Investment*, 1–28.

La Torre, M., & Chiappini, H. (2020). Sustainable finance: Trends, opportunities and risks. In *Contemporary issues in sustainable finance* (pp. 281–287). Palgrave Macmillan.

Migliorelli, M., & Dessertine, P. (2019). *The rise of green finance in Europe: Opportunities and challenges for issuers, investors and marketplaces*. Palgrave Macmillan.

Park, S. K. (2018). Investors as regulators: Green bonds and the governance challenges of the sustainable finance revolution. *Stanford Journal of International Law, 54*, 1.

Quatrini, S. (2021). Challenges and opportunities to scale up sustainable finance after the COVID-19 crisis: Lessons and promising innovations from science and practice. *Ecosystem Services, 48*, 101240.

Semieniuk, G., Campiglio, E., Mercure, J. F., Volz, U., & Edwards, N. R. (2021). Low-carbon transition risks for finance. *Wiley Interdisciplinary Reviews: Climate Change, 12*(1), e678.

Wang, Y., & Zhi, Q. (2016). The role of green finance in environmental protection: Two aspects of market mechanism and policies. *Energy Procedia, 104*, 311–316.

Weber, O. (2016). Finance and sustainability. In *Sustainability science* (pp. 119–127). Springer.

Zhang, D., Zhang, Z., & Managi, S. (2019). A bibliometric analysis on green finance: Current status, development, and future directions. *Finance Research Letters, 29*, 425–430.

CHAPTER 2

What's in a Name? Mapping the Galaxy of Green Finance

2.1 Introduction

Sustainable economic and financial development calls for green finance, as a commonly used strategy for dealing with environmental problems. The need for action to reconcile economic development with environmental protection, and to achieve objectives defined as the common good, has long been recognized by scholars, as well as by the international community (Ekins, 2002; Gibbs, 2003; United Nations, 2016; Xepapadeas, 2005). The 1987 World Report Commission on Environment and Development (Brundtland Report) was followed by numerous initiatives, culminating in 2015 in the approval of the United Nations' 2030 Agenda for Sustainable Development and its 17 objectives (Sustainable Development Goals—SDGs), divided into 169 targets to be achieved by 2030 (Gupta & Vegelin, 2016). In December of the same year, after the Paris climate conference, 195 countries joined the first universal and legally binding agreement, identified with the acronym COP21 (Rhodes, 2016). All the participants agreed on a plan of global action, with the aim of preventing the effects of climate change, and to contain global warming below 2 °C compared to pre-industrial levels (Sweet, 2016). After COP21, most countries that produced the highest gas emissions planned further actions to help bring their economy to zero emissions by 2050 (Jaumotte et al., 2021).

© The Author(s), under exclusive license to Springer Nature Switzerland AG 2022
A. Rizzello, *Green Investing*, Palgrave Studies in Impact Finance, https://doi.org/10.1007/978-3-031-08031-9_2

In such a context, the transition to an environmentally friendly economy—also referred to as the *low-carbon* economy (Bridge et al., 2013), *net-zero* economy (Bonsu, 2020), *circular* economy (Stahel, 2016), or *green* economy (Loiseau et al., 2016)—required higher financial resources than the public sector budgets (Lagoarde-Segot, 2020). The bridging of this *funding gap*, estimated at an immediate requirement of USD 1.5 trillion annually until 2030 (Doumbia & Lauridsen, 2019), requires developing financial tools and completely re-examining the financial theory and practice, by providing a new direction for the innovation of traditional financial theory. As a result, green finance came into being, by offering new profit opportunities for the financial industry, while also contributing to a positive environmental impact (Berger, 2011).

However, it is at the European level that the green-targeted public policy initiatives regarding financial market regulation, public finances, and monetary policy have become more prominent (Bongardt & Torres, 2022). European efforts to fight negative environmental impacts started in 2001, with the Strategy for Sustainable Development, and were then integrated into the European 2020 strategy. In 2019, with the launch of the *Green Deal*,[1] the European Union (EU) planned to contribute to carbon neutrality by 2050 through mobilizing up to one thousand billion euros in the next ten years (Krämer, 2020). More recently, the European Parliament has presented a taxonomy, elaborated by a Technical Expert Group, to provide a framework for green finance to more effectively reach the *European Green Deal* targets (Brühl, 2021). At the end of 2021, the extraordinary funding program *Next Generation EU* (NGEU), adopted to counteract the negative economic effects of the COVID-19 pandemic, confirmed the prioritization of the green transition in European policy; it affirmed that the financed projects would not have negative environmental effects (De la Porte & Jensen, 2021).

According to Global Sustainable Investment Alliance report in 2020, global sustainable investment reached USD 35.3 trillion in five major markets, a 15% increase since 2018; the United States and Europe continued to represent more than 80% of global sustainable investing assets. However, the enormous volume of investment is very hard to

---

[1] The European Green Deal is the EU's strategy for reaching its 2050 climate goal. Specifically, it is a package of policy initiatives, which aims to set the EU on the path to a green transition, with the ultimate aim of achieving climate neutrality by 2050. For further details: A European Green Deal | European Commission (europa.eu).

quantify due to the global, multisectoral nature of the problem, and the lack of reliable data (Van Steenis, 2019).

The aim of this chapter is to provide an overview of the complexity of green finance. For this purpose, Sect. 2.2 precisely defines green finance, and Sect. 2.3 identifies its conceptual boundaries and nature. The market ecosystem is analyzed in Sect. 2.4, by highlighting the asset classes and typologies of financial instruments used in the green finance. Section 2.5 discusses the issue of the standards and frameworks of environmental investing. Section 2.6 presents an overview of the size and segments of the green finance market, and Sect. 2.7 concludes.

2.2 Definition and Key Features

Green finance is a relatively new but clearly identified field of finance research. To date, practitioners and scholars have failed to establish a commonly accepted definition of the term (Lindenberg, 2014). In particular, many publications fail to define green finance, and those that do so use a variety of workable definitions (Berensmann & Lindenberg, 2019). Furthermore, across the financial sector, diverse approaches and definitions are often developed separately, which vary in their scope, level of detail, transparency, and other dimensions (Inderst et al., 2012). Such definitional opacity makes it difficult to assess the overall progress of the green finance market (Sachs et al., 2019).

The beginning of the green finance concept can be traced to the launch of The United Nations Environment Program Finance Initiative (UNEP FI) in 1992, when UNEP formed a partnership with a group of global financial actors, with the aim of mobilizing private sector finance for sustainable development (UNEP FI, 1992) by integrating environmental considerations into financial services and practices. Two years later, after the implementation of the international environmental treaty (United Nations Convention on Climate Change), which was adopted at the Rio Earth Summit in June 1992, the first annual UNEP FI Global Roundtable was held in Switzerland, on the topic of Greening Financial Markets (Mendez & Houghton, 2020). In the following years, two academic publications (Devas, 1994; Kublicki, 1993) for the first time used the term "green finance" in their title. While the first centered on green finance as an alternative solution to the debt of lesser developed countries, the second focused on the role of green finance—described as an apparent paradox due to the juxtaposition of "green" and "finance"—in helping to solve the global problem of achieving sustainable development.

Since 1994, a growing number of international policy initiatives on environmental protection were accompanied by a collateral setting of environmentally oriented financial standards and principles. For instance, at the Millennium Summit in September 2000, the world leaders ratified the United Nations Millennium Declaration, which set out eight goals for developing countries with specific targets for 2015, known as the Millennium Development Goals (MDGs). In 2003, the Equator Principles (EPs) were adopted by a group of international banks, with the aim of providing a common baseline and risk management framework for financial institutions, regarding environmental and social factors when financing projects (Schepers, 2011). Three years later, in 2006, over 1,500 financial actors signed the Principles for Responsible Investment (PRI), and committed to incorporating Environmental, Social and Governance (ESG) issues into investment practice (Gond & Piani, 2013). The same time period (1992–2011) can be considered the early stage of research on green finance, with a small number of publications that mainly described the establishment of a basic system of green finance (Labatt & White, 2002).

A second period, from 2012 to 2017, is characterized by a stable development of the field. This includes the three main policy milestones that assisted the recent growth of the green finance sector:

- the Paris Agreement in 2015 (or COP21—the 21st Conference of the Parties) refers to the treaty signed by 195 countries to combat climate change and begin actions and investment toward a low-carbon, resilient, and sustainable future);
- The 2030 Agenda for Sustainable Development, adopted by world leaders in September 2015, was followed in January 2016 by the definition of the 17 Sustainable Development Goals (SDGs), a set of targets that mobilized all countries' efforts to end all forms of poverty, fight inequalities, and tackle climate change;

- The Capital Markets Union[2] accelerated reforms, issued by the European Commission in September 2016, stating that reforms for sustainable finance are necessary to ensure the building of a sustainable financial system, and contribute to creating a low-carbon, climate-resilient economy.

During this period, the research on green finance increased significantly, and different contributions covered various research streams and topics. However, many publications did not provide a common definition of green finance, except in the grey literature. For Zadek and Flynn (2013), green finance includes all capitals directed toward the transition to a green economy, and is interchangeably used with "green investments." For Höhne et al. (2012), the term refers to financial investment in sustainable development or environmental products, policies, and projects; while for Pricewaterhouse Coopers Consultants (2013), the expression means those financial products and services that promote environmentally responsible investments.

The first attempt to provide a holistic definition of green finance was made by Lindenberg (2014), with a tridimensional concept to explain its complex nature. In particular, Lindenberg's concept of green finance includes (*i*) a component related to the financing of public and private green investments, (*ii*) the financing of public policies in environmental projects and initiatives, and (*iii*) a component related to the financial system.

For the G20 Green Finance Study Group (2016), green finance is the financing of investments that provide environmental benefits in the broader context of environmentally sustainable development. The Organization for Economic Cooperation and Development (OECD) (2017) considers green finance to be a strategy for achieving economic growth while ensuring the reduction of environmental damage. The European Banking Federation's (EBF) definition of green finance (2017) includes

---

[2] The Capital Market Union (CMU) is an economic policy initiative launched in 2016, which aims to foster stronger, sustainable economic growth by creating deeper and more integrated capital markets in the European Union. The initiative was renewed in September 2020 by the European Commission, which issued a new CMU action plan formed of 16 legislative and non-legislative measures to support green and long-term investments. For further detail: What is the capital markets union? | European Commission (europa.eu).

both environmental and climate change-related aspects, but it is not limited to those two elements.

The third period, 2017 to 2022, has featured vigorous research on green finance, and includes the publication of the most influential academic articles in this field. It is important to note that in the same period, further policy and societal milestones influenced scholars' growing interest in the topic of green finance:

- EU Regulation 2020/852 was introduced to the European regulatory system; it included a taxonomy of economic activities, which can qualify as "sustainable" based on their alignment to the EU's environmental objectives, and their compliance with a number of social clauses. The novel aspect, within the broader policy target of the European Commission's (EC) Action Plan on Sustainable Finance (adopted in March 2018), was also directed at investors, in terms of integrating sustainability themes into their investment policies, and understanding the environmental impact of the economic activities in which they invest or might invest (among others, Ahlström & Monciardini, 2021);
- The emergence of the global youth-led and -organized global climate strike movement *FridaysForFuture*, which significantly increased the centrality of environmental issues in the global policy agenda (among others, Kaneko, 2021);
- The growing financial market of sustainable finance, in different and innovative financial approaches (for an overview, see Lehner, 2016; La Torre & Chiappini, 2020);
- The negative effect of the COVID-19 spread, which produced a decisive shift toward the transition to sustainable and climate-neutral business models (among others, Galanakis et al., 2022)

In these years, the definition of green finance significantly reduced its emphasis on the broader aspects of the greening of the financial system (Falcone et al., 2018; Ng, 2018), or on the financing of green policies interventions (D'Orazio & Popoyan, 2019; Taghizadeh-Hesary & Yoshino, 2019). It focused more on the aspect of capital allocation and investment (*i*) by private capitals (Clark et al., 2018; Taghizadeh-Hesary & Yoshino, 2019); (*ii*) through the financial system (Fu & Ng, 2020; Weber & ElAlfy, 2019; Ziolo et al., 2019); (*iii*) in particular, through banks (Cui et al., 2018; Park & Kim, 2020; Yuan & Gallagher, 2018; Zou et al., 2020).

Briefly, from the analysis of the green finance literature, the thinking about green finance has progressed through different stages over the last three decades. The focus is gradually shifting from macro considerations concerning the green financing policies and investment for development (Meyer, 1997; Wang & Zhi, 2016), toward long-term environmental value creation (Kurznack et al., 2021) through financial instruments (Fig. 2.1).

Thus, different though these definitions are, they all share common elements. In more detail, green finance refers to:

- the allocation of capital for a wider purpose than the financial alone;
- the focus of the investment is the (direct or indirect) provision of benefit for the environment, or reducing harm to it;
- the management of environmental risks for financial services and products;
- the framing of policies and infrastructures to enable the transition;
- a subset of a financial system oriented to supporting sustainable development.

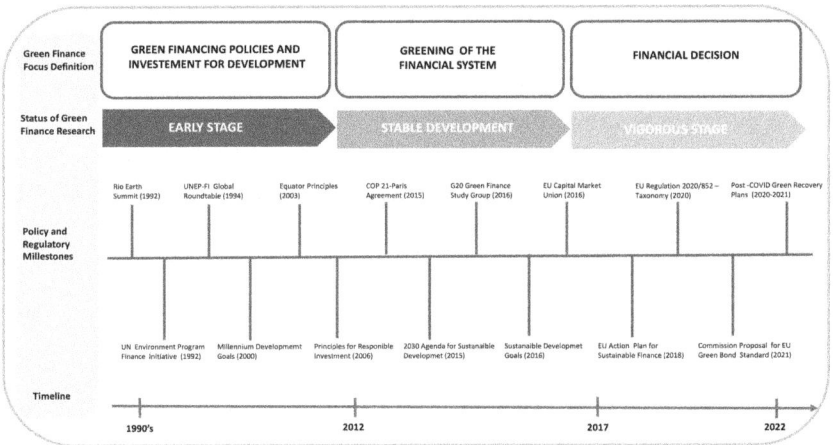

Fig. 2.1 Parallel evolution of green finance policies, green finance research and green finance definitions' focus (*Source* Author's elaboration)

To conclude, a complete definition could consider green finance as:

> a subset of a sustainable financial system that enables the allocation of capital toward policies, organizations, investments, and activities which provide direct/indirect environmental benefits, intended both as lower negative impact on the environment than the status quo, or as positive environmental impact.

2.3 The Theoretical Boundaries of Green Finance

The evolution in the definitional elements of green finance cannot be exhaustive enough to capture the complexity of this field. The conceptual elements of a complete definition of green finance represent just one piece of the green finance mosaic. Indeed, most of the recent contributions on this topic (Akomea-Frimpong et al., 2021; Berrou et al., 2019) highlight a certain richness of (i) interchangeable and (ii) sectorial terms. This phenomenon originates from many reasons, mainly attributable to:

- the variety of different approaches to defining the activities related to the "finance for green returns," adopted from the multitude of policy initiatives worldwide;
- the presence of a certain number of "motor themes" among green activities (climate goals or renewable energy sources, among others);
- the use of "niche" terms adopted by organizations and scholars who approach the green theme without having a financial background.

In particular, it is possible to observe the use of the following overlapping green finance terms:

- Environmental Finance;
- Environmental Investing;
- Sustainable Finance;
- Sustainable Investing;
- ESG (Environmental, Social and Governance) Investing.

The identification of sub-sectorial terms for green finance was not easy, and required intense activity in the triangulated content analysis of: (i) the scientific publications extracted from the field, (*ii*) the research articles that cited those publications, and (*iii*) the reports of the main international organizations active in the green finance sector. The following sub-sectors were identified:

- Climate Finance
- Mitigation/Adaptation Finance;
- (Low-)Carbon Finance;
- Blue Finance;
- Biodiversity Finance;
- Conservation Finance;
- Forest Finance;
- Circular Economy Finance;
- Desertification Finance;
- Renewable Energy Finance;
- Water, Sanitation, and Hygiene Finance;
- Green Building Finance.

Thus, the boundaries of green finance can be analyzed by considering:

- the relationship between green finance and other interchangeable terms (section "Green Finance and Its Alter Ego"), and
- what the term "green" means for green investors (section "The Green Galaxy of Green Finance").

Green Finance and Its Alter Ego

"Environmental finance" and "environmental investing" represent the most famous, and probably the oldest, terms used to refer interchangeably to green finance. However, based on analyzing the literature, the two terms appear not to perfectly overlap, even if their roots are essentially similar. Indeed, in contributions referring to environmental finance, it is observed that green finance shares the same attention to the need to transition to a sustainable economy through a sustainable financial system, but with a focus on the financial implications of environmental changes for industry and firms (Anderson, 2016; Labatt & White, 2002). Specifically, two main features characterize environmental finance's conceptual structure. First, regulatory solutions to encourage investment in new environmentally friendly technologies and solutions, such as clean energy infrastructure (Hafner et al., 2019; Jones, 2015). Second, how environmental changes impact corporate assets (Dietz et al., 2016) and the related adaptation investments (Mekonnen, 2014).

The family of green finance alternative terms has been extended more recently with two new members: *sustainable finance/investing* and *ESG investing*. While some organizations use these terms interchangeably, green, sustainable finance, and ESG investing are nested concepts. The reasons for this recent marriage are twofold. First, the term "sustainable finance," overall in the EU, has overlapped with the financing of environment-related activities, especially after the definition of the Sustainable Action Plan and related regulation drafts. For instance, even if the actions in the plan focused on both environmental and societal outcomes, the environmental aspects remained central in the main regulatory documents. Only at the end of 2021 were documents produced that related to social activities, such as the Social Taxonomy (European Union Platform on Sustainable Finance, 2021). Thus, from 2018 to 2021, the debate in European countries on the term "sustainable finance" essentially referred to a number of policy and regulatory actions that were environment-centered (Eckert & Kovalevska, 2021).

Furthermore, a second reason for this ambiguity derives from the lack of lexical rigor from various actors that combine sustainability/green labels with green activities and products (Cho & Taylor, 2020). Similar considerations apply to ESG investing, one of the most well-known and debated forms of sustainable investing, which focuses on different non-financial dimensions of a stock's performance: Environmental, Social, and Governance (Van Duuren et al., 2016). Even if ESG investing includes social as well as governance dimensions, along with the public sector initiatives to fulfill the Paris Climate Agreement, there has been growing recognition that encouraging investors' use of ESG as a climate-aligned approach will play a critical role in supporting an orderly transition to low-carbon economies (OECD, 2017).

Despite progress, challenges remain, such as in the promulgation of frameworks, data inconsistencies, lack of comparability of ESG criteria and rating methodologies, as well as inadequate international disclosure standards regarding how ESG integration affects asset allocation (Boffo &

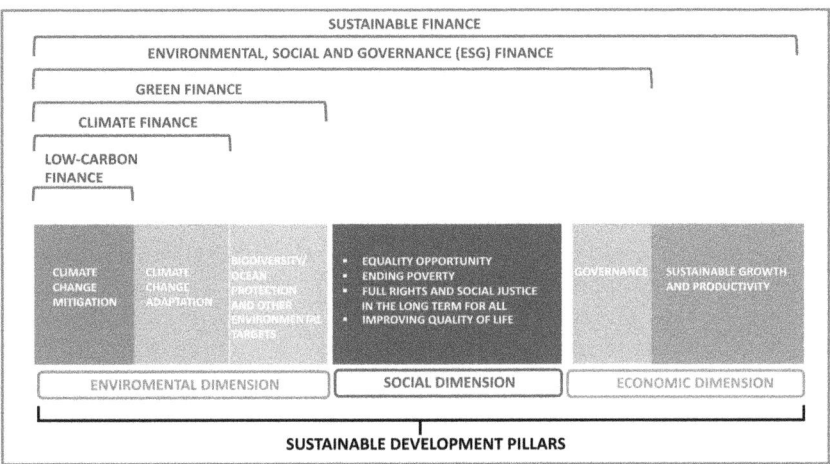

Fig. 2.2 Green finance and other forms of finance for sustainability (*Source* Author's elaboration)

Patalano, 2020). As in the case of sustainable finance, the ESG environmental goals' focus also covers only a part of the entire spectrum of sustainable activities, as explained in Fig. 2.2.

The analysis of the overlapping of these definitions illustrates that the term "green finance" focuses on what it finances (for instance, investment in green companies), and not on the specific environmental benefit that the investment is intended to achieve. However, the recent evolution of definitions presents green finance not as an objective in itself, but rather as a tool to improve environmental conditions, by focusing on the potential impact of green investments. Thus, this overview provides a foundation for the next chapters, and enables an insightful investigation into the green finance puzzle.

The Green Galaxy of Green Finance

The adoption of "green finance" niche terms has been identified in many studies and reports. All the terms relate to sub-sectors of green activities or the green business sector, such as climate, forestry, waste management, water, sea and ocean, or renewable energy.

Climate finance refers to the use of financial resources to sustain the cost of transitioning to a low-carbon global economy, and to build resilience against climate change impacts. According to the content of the extracted publications, climate finance refers overall to macro-economic considerations regarding financial flows directed to fight climate change, including public expenditure (Buchner et al., 2011), public sector investments (Nakhooda et al., 2015), as well donor funding or development aid that is consistent with the principles of the United Nations Framework Convention on Climate Change (UNFCCC) (World Bank, 2017).

International policy actions have distinguished two options for addressing climate change, one in the long term, i.e., mitigation (intended as various actions for reducing the sources or enhancing the sinks of greenhouse gases); and adaptation (responding at a local level, generally in the short term, to the effects of climate change) (Biesbroek et al., 2009; Jordan et al., 2010). Consequently, there is also a division between adaptation and mitigation in the financial resources mobilized to cope with climate change (Illman et al., 2013). In the related literature, the motor topics are essentially the same as those recorded in climate finance-related articles, with the dominant theme being the financing of public policies for adaptation and mitigation.

In addition, the sub-field of carbon finance is related to the achievement of climate targets. However, it is distinguished from climate finance, because its central activity concerns the de-carbonization of economic activities. For these reasons, carbon finance presents multiple streams, ranging from markets that price and trade emissions rights (Bryant, 2019; Louche et al., 2019), to raising capital expressly for low-carbon investment in enterprises, projects, and initiatives (Clapp & Stephens, 2019; Sullivan, 2018).

A further motor theme, within the area of reducing gas emissions on the one hand, and in the protection of natural resources on the other, relates to the field of the renewable energy sector. Renewable energy has a long history, deriving from the modern energy economy, which is dominated by fossil fuels. Most actions for significantly reducing carbon emissions rely heavily on these renewable energy sources. This has involved a large amount of public and private investments in renewable energy projects over the last two decades (Mazzucato & Semieniuk, 2018).

The importance of protecting and preserving open lands and the ecosystem is emerging everywhere. The related financial aspect is commonly indicated with the term "conservation finance" (Clark, 2012); this can be identified as a means of delivering maximum conservation impacts, while at the same time generating returns for investors (Kay, 2018). More recently, the theme was further segmented with the rise of biodiversity finance (Meyers et al., 2020), marine conservation finance (Bos et al., 2015), and forest finance (Liagre et al., 2015). More recently, to improve the drinking water and sanitation-related standards, specific targets have been defined under the Sustainable Development Goals (SDGs). Water, sanitation, and hygiene finance aims at reducing pollution caused by untreated sewage and poorly managed fecal sludge, and increasing the reuse of treated wastewater, by making it financially feasible to implement the related goals and targets (Hutton & Varughese, 2016).

The transition toward a circular economy also represents an emergent frontier for green financial actors (Goovaerts & Verbeek, 2018). "Circular economy" is an innovative green sector, which promotes the efficient use of resources through waste minimization, long-term value retention, and reduction of primary resources use (Morseletto, 2020). The financing of activities and organizations based on a circular business model represents the financial aspect of this green sector.

In order to provide an intuitive representation of green finance and its satellite terms, a visual representation (Fig. 2.3) is provided by matching each term with the classical elements of nature: earth, water, air, and fire.

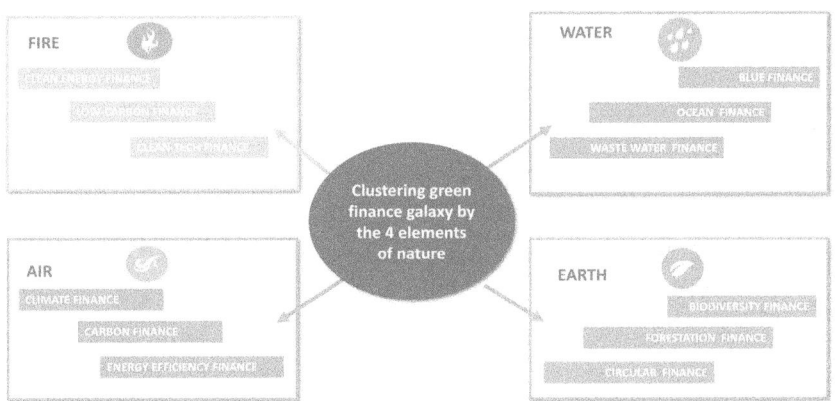

Fig. 2.3 Green finance definition's galaxy: a four-element visualization (*Source* Author's elaboration)

2.4 The Green Finance Ecosystem

The green finance ecosystem comprises at least four components: the demand for capital, the supply of capital, the intermediaries and channels that connect the two, and the broader context within which all actors operate. The *supply side* includes all those financial actors that are actively making investment decisions, using various forms of finance, which are deployed through diverse forms of finance.. The *demand side* includes a range of organizations that are organized with capital and work to generate financial returns and environmental benefit, in response to green investment needs expressed by a number of green-seeking actors. All these actors typically operate within wider enabling conditions which convert green investment "needs" into actual demand.

Furthermore, markets operate according to an evolving set of political, social, and cultural norms and conventions. Figure 2.4 represents a synthesis of this ecosystem.

The Demand Side

The segmentation of the demand side of a green finance ecosystem includes those actors seeking environmental outcomes (Owen et al.,

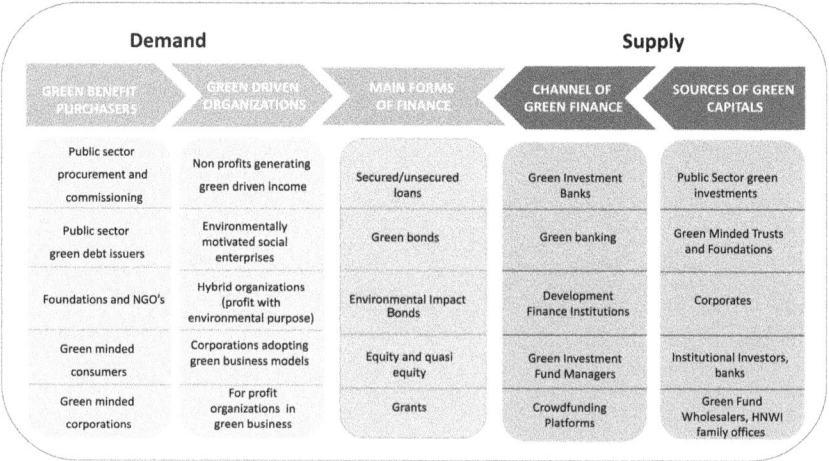

Fig. 2.4 The green finance ecosystem (*Source* Author's elaboration)

2018), and others that can be defined as environmentally driven organizations (Albino et al., 2009). In the first category, it is possible to include public sector demand for environmental outcomes pursued by procurement, as well as by commissioning environment-focused programs. More recently, governments have started to match their green objectives with financial markets' demand for green capitals by issuing sovereign green bonds (Schmittmann & Chua, 2021). Other actors engaged in commissioning environmental outcomes are green-focused foundations, or international non-governmental organizations active in this field. Green-minded consumers (Luzio & Lemke, 2013) and green-minded corporations' (Min & Galle, 2001) represent a further important area of green outcome demand.

Green-driven organizations can include various forms of entrepreneurial entities (Walley & Taylor, 2002), ranging from non-profits with green-driven income-generating activities, to environmentally motivated social enterprises (Vickers & Lyon, 2014), or hybrid organizations that adopt environmental objectives (like non-profits), but generate income to accomplish their mission (like for-profits) (Haigh & Hoffman, 2011). Furthermore, corporations that adopt green-oriented CSR activities (Babiak & Trendafilova, 2011) can also be included within the demand side of this scheme. Finally, traditional for-profit organizations that operate in green business sectors, or are based on green business models (Trapp & Kanbach, 2021), represent the biggest growth segment of the ecosystem's demand side.

The Supply Side

The supply side of the green investment ecosystem can be framed as actors that generate environmentally oriented capitals, and the channels of such capitals. The first category includes:

- Asset owners, such as public sector/governmental investments, green-minded trusts and foundations, corporations, high net worth individuals (HNWIs), retail investors, and asset managers;
- Asset managers, which include green funds wholesalers, family offices, institutional investors, and banks.

Among the mentioned actors, banks, financial institutions, and funds are expanding their importance within the green finance ecosystem. This is due to the building and improvements, especially in the EU area, of green investment regulatory frameworks, which incentivize their role in the green financial system (Hafner et al., 2020). Within this context, the role of such financial actors is growing rapidly, especially with regard to green bond issuing activities.

The channels for opening the market to green capitals are multifaceted. Beyond the rise and evolution of green finance in the banking sector context (Akomea-Frimpong et al., 2021), it is also possible to note the importance of other forms of financial intermediaries, which include:

- *Green Investment Banks* (Schub, 2015): Public or quasi-public institutions that use limited public capital to leverage greater private investment in green activities, especially in renewable energy (OECD and Bloomberg Philanthropies, 2015);
- *Development Finance Institutions* (Fuchs et al., 2021) play a key role in achieving the environmental and climate goals of aligning financial flows with low-emission, climate-resilient development pathways; primarily focused on direct project financing.
- *Alternative Banks*, also identified as ethical banks or social banks (Weber & Remer, 2011). Their focus is not on maximizing profits for shareholders, but on maximizing environmental value for the wide audience of stakeholders, going beyond pure profits.

Asset management has become one of the fastest-growing segments in the green financial industry. Asset managers are likely to specialize in the areas of advisory or discretionary management on behalf of investors; these services normally require rigorous financial analyses, combined with asset and stock selection, plan implementation, and regular monitoring and reporting of investment activity. The principal sectors of the green asset management industry include mutual funds, pension funds, and private–client assets (Lesser et al., 2014). On the other hand, green investment fund managers include different sub-segments, such as fossil-free fund managers or climate fund managers, depending on whether their business models are based on a specific green sector (Bracking, 2015; Collinson, 2015). Green investment funds only invest in companies that they consider to be, in some way, aligned to environmentally responsible activities (Scholtens, 2011). Green funds typically allocate their capitals to a wide range of companies across different industries, which helps to limit the risk to investors (Chang et al., 2012).

More recently, crowdfunding platforms have also actively registered a growth (Adhami et al., 2017; Cummings et al., 2017; Lam & Law, 2016) within the green finance ecosystem. Crowdfunding refers to mobilizing a multitude of single investors (the "crowd"), on dedicated websites known as crowdfunding platforms, to finance projects that are, in the case of green finance, environmentally oriented projects (Nigam et al., 2018).

The Forms of Green Capital and Related Instruments

Fischer (2014) divides the main forms of green finance into four categories:

- Debt, which includes, for example, secured loans, unsecured loans, and green bonds;
- Hybrid, such as environmental impact bonds;
- Equity and quasi-equity;
- Grants.

Equity financing, often used in the early stages of developing a project or company, involves acquiring a permanent share (also defined as "stock") in the ownership of a corporation, or an ownership interest in a project. Equity can be generally split into preferred stock, and ordinary or common stock. Equity finance can be received through two principal channels, each of which involves different actors: these are privately raised funding and publicly raised funding.

The first involves "direct" acquisitions of green equity capital by a company or project; rather than through a stock exchange, as in the second case. Private injection of green capital into a company is given different names: "green angel capital," "green venture capital," and "green private equity capital" (Liu & Keleher, 2009; Randjelovic et al., 2003). Venture capitalism is a more structured form of angel investing, performed by groups of wealthy individuals and specialized teams of venture capitalists, who target returns that are multiples of their original investment. Private equity finance is provided from pooled funds to companies at a mature stage of development; thus, they involve lower risks, and the expected return is also likely to be lower than in the previous cases. Second, in publicly raised funding, equity is raised through an Initial Public Offering (IPO) or capital captured through stock markets. The funded companies in that case tend to be large corporations, with significant balance sheets.

The debt-based green finance includes the promise to repay the capital sum and the predetermined return that distinguishes debt from equity. As with equity, debt can travel through different channels, privately and publicly, in public markets. In the first case, the main actors are represented by commercial banks, which provide the majority of debt funding privately through loans. In the second case, an institution (such as a government or large company) issues a bond in the open market with a specified principal, a term, and a specific coupon payment. In the case of publicly raised bonds, the main actors are determined by the type of institution issuing the bond, ranging from a company (corporate bond), to national government (sovereign bond) or local authority (municipal bond). Finally, grants are typically provided for non-revenue generating activities, such as capacity-building programs or ongoing activities that do not generate financial return. Generally, the main source of grants for environmental action has been international financial institutions, bilateral institutions, and international climate funds. Grants are often released in combination with debt capital (Martin, 2015; Sonntag-O'Brien & Usher, 2012).

Among debt-based green finance instruments, loans represent the lower-developed instruments (Berrou et al., 2019), but show a significant rate of growth (Degryse et al., 2021). Within the context of green finance, loans received a standard framework by the Loan Market Association (LMA), together with the Asia Pacific Loan Market Association (APLMA) and Loan Syndications and Trading Association (LSTA). They issued a framework of market standards and guidelines, the Green Loan Principles, to promote the development and integrity of the green loan products (Li et al., 2018). For these purposes, green loans can be any type of loan instrument directed to financing or re-financing, in whole or in part, new and/or existing eligible green projects; they are based on the following four core components:

- Use of Proceeds;
- Process for Project Evaluation and Selection;
- Management of Proceeds;
- Reporting.

In addition, in order to be labeled as green loans, all designated loan activities for green projects should provide clear environmental results, which will be assessed, and possibly, quantified, measured, and reported by the borrower; they therefore incur additional organizational and operational costs (Giraudet et al., 2021), which represent a barrier to the mainstream use of this instrument. However, the loans' diffusion is expanding among banks (Umar et al., 2021) with the creation of innovative green lending products, such as:

- Green Mortgages (e.g., to buy energy-efficient homes);
- Commercial Building Loans (e.g., loans for new efficient condos);
- Home Equity Loans (e.g., to offer financing options for purchasing and installing residential solar technology);
- Fleet Loans (e.g., to truck companies, to finance fuel-efficient technologies).

Green bonds are fixed-income securities which are structured with the same characteristics as standard bonds in terms of seniority, rating, execution process, and pricing, but with proceeds dedicated to climate or environmental sustainability purposes (Tang & Zhang, 2020). The European Investment Bank (Banga, 2019) issued the first green bond, targeted to climate benefits, in June 2007.

As in the case of loans, the issuing of green bonds has been standardized by the International Capital Market Association, which provided a set of voluntary process guidelines for issuers (ICMA, 2021), called the Green Bond Principles (GBP). The GBP aim to promote integrity in the development of the green bond market, but fail to state-specific environmental impact targets, or impose limits on the categories of projects and activities that can be financed by such instruments. As in the case of green loans, there are four main components of a green bond issuing process, as established by the GBP:

- *Use of Proceeds*, by explicitly recognizing several broad categories of potential eligible green projects, such as renewable energy, energy efficiency, sustainable waste management, or sustainable forestry and agriculture;
- *Process for Project Evaluation and Selection*, to determine which projects will be funded.

- *Management of Proceeds* addresses transparency in tracking the proceeds in the allocation of green bond to a specific sub-portfolio;
- *Reporting* addresses the disclosure on use of proceeds, project descriptions, and expected environmental impact.

However, if the ICMA's Green Bond Principles represent the most widely agreed voluntary standards and guidelines for issuers, several other guidelines and regulations have been issued worldwide both at an international or national level. The first group includes the Climate Bond Standards developed by the Climate Bonds Initiative (CBI), a voluntary disclosure framework that adds more specific criteria concerning climate-eligible activities to those incorporated in the Green Bond Principles, and serves as a screening tool for investors and governments. The national frameworks include:

- *Green Financial Bond Guidelines*[3] (China), developed by the People's Bank of China (PBOC) and China Society of Banking and Finance; they are aligned with GBP and CBI's standards and include mandatory quarterly reporting for issuers;
- *Green Bond requirements*[4] (India), published by the Securities and Exchange Board of India (SEBI); they are in line with GBP and include further recommendations and requirements, as indicated in India's Nationally Determined Contribution to the Paris Agreement;
- *Transition Energetique Climat' label*[5] (France), a mandatory requirement for green funds to invest in green bonds aligned with the GBP and CBI Taxonomy, in order to be recognized with the label;
- Stock Exchanges' listing requirements for green bonds (London, Shanghai, Shenzhen, and Luxembourg); these concern the development of minimum requirements for listing green bonds.

---

[3] Full information available at: http://www.pbc.gov.cn/english/130721/3131759/index.html.

[4] Further information available at: Microsoft Word—Board Memo-Agenda no 4—Disclosure requirements for issuance and listing of Green Bonds (sebi.gov.in).

[5] Further information available at: Le label Greenfin | Ministère de la Transition écologique (ecologie.gouv.fr).

In the fast-growing green bond universe, some different types can be identified; in particular:

- *Climate Bonds* focus in particular on projects related to climate change mitigation and adaptation; they represent the most mature and well-distinguished product among the green bond labels;
- *Green Revenue Bonds*: a non-recourse-to-the-issuer debt obligation in which the credit exposure in the bond is to the pledged cash flows of the revenue streams, fees, or taxes;
- *Green Securitized Bonds*: built on the basis of the traditional scheme of securitization. In this model, the bond-holding vehicle purchases portfolios of debt secured against renewable energy projects from banks. Such products can be considered as being "green" when the issuance of the securities is in line with the criteria for labeling a green bond.

The green private equity industry can be broken down into different segments. In particular, Green venture capital firms usually invest in younger companies that adopt innovative green tech solutions, as minority shareholders. Buyouts target larger companies in more mature growth stages, and provide growth capital. Leveraged Buyouts (LBO) are specific deals in which a small share of equity is invested and leveraged by a large acquisition debt. With regard to equity instruments, increasing consideration is being paid to Initial Public Offerings (IPOs) from clean technology providers, carbon credit developers, and other firms marketing environmental products and services to accredited investors; these include angel investors/venture capitalists, high net worth individuals, and early investors (Knight, 2011). For instance, in 2018, JP Morgan supported its market debut with a USD 270 million IPO of the clean energy "unicorn" Bloom, which originated from a NASA project on Mars (Waters, 2018). Finally, the evolution of the green finance sector has also largely involved equity funds, which has led to increasing complexity in the assessment of green investment eligibility. For instance, first-generation green funds solely employ exclusionary environmental criteria; second-generation funds use positive criteria that concentrate on progressive environmental policies and practices; and third-generation funds apply both exclusionary and positive criteria to assess and select potential investments. Within such context, indices are a primary investment tool for investment managers and investment owners, as they provide a benchmark or point of reference for the active investment decisions. Among major index providers, there is a wide choice of equity indices available, which use different approaches, definitions, and composition (Petry et al., 2021). Furthermore, some indices have a sectoral focus, such as on clean technology, or on a prominent factor, such as carbon emissions; while others span the entire range of green activities. The preferences for indices differ across countries and investors, and three main approaches have been identified for selecting and weighting the index constituents: screening, best-in-class, and reweighting (Rezec & Scholtens, 2017). There are also major differences in the metrics used to select green stocks. As a result, some indices favor investment in specialist green companies, while others are designed to select the best companies within a sector.

In terms of hybrid instruments, it is possible to observe the rise of the Environmental Impact Bond (EIB), an innovative instrument of the impact investing industry; this adopts a positive financial approach to intentionally obtaining financial returns alongside social and/or environmental returns, expressed in the form of measurable positive impact (Chiappini, 2017). The EIB, a subset of the broader expression Social Impact Bond (SIB), is the financing, based on a pay-for-success logic, of a predetermined environmental project with a predetermined environmental impact. The EIB involves investors, the project delivery organization, and the commissioner, which is usually a governmental entity that repays investors at the end of the program only in case of success (intended as the achievement of environmental benefit above the predetermined threshold). If the program fails to achieve success, investors will lose their principal. Hence, EIBs, like all SIBs, are conceived as hybrid instruments, incorporating both debt and equity logics. The adoption of such innovative instruments presents huge potential, in terms of enhancing the profitable options for using private capital to fund environmental outcomes.

Finally, of particular importance are green finance products of banks, which play a central role in financing green projects. The green finance products from banks (Buchner et al., 2011; He et al., 2019; Wang et al., 2017; Yuan et al., 2020) include:

- green loans (credit);
- green long-term investment accounts;
- carbon finance;

- climate finance;
- green traded stocks and bonds;
- green bancassurance;
- green infrastructural finance.

Though many of these terms have been sufficiently described above, it is useful to briefly define green long-term investment accounts, green bank assurance, and green infrastructural finance. The first relates to a form of innovative products from banks that allow customers to deposit funds to support environment-oriented long-term activities (He et al., 2019). Green bancassurance offers coverage for green buildings and other insurable green assets (Wang et al., 2017). Finally, green infrastructural finance refers to support for ecological projects, in building large infrastructural interventions that provide long-term environmental benefits.

2.5 The Green Finance Standards and Frameworks

Due to the complexity and rapid evolution of the green finance market, several efforts of harmonization are required to support all the stakeholders involved in building a green financial system that can channel finance and investments into all the ecologically friendly activities and projects. Consequently, to enable the financial sector to support the financialization of the green transition, a number of actions have been developed over the last two decades (Heaps & Guyatt, 2017). According to Green Finance Platform (2021), more than 680 policy and regulatory measures have been adopted worldwide, with a percentage of increase of 264% in the last five years.

For instance, it is possible to identify green finance codes, taxonomies, guidelines, and catalogues for green finance. Also commonly defined as "standards" or "frameworks" (Herbertson & Hunter, 2006; Nedopil et al., 2021), they have involved a proliferation of competition initiatives that are difficult to harmonize. The main international green finance standard-setters are governments, followed by public and private finance institutions. Also active in this field are non-financial and non-government institutions, associations, and think tanks. The spectrum of green finance standard setters is represented in Table 2.1.

Table 2.1 The green finance standard setters: an overview

Typology	Name of issuer (main list)	Standard/framework outputs (main list)
Governments and national/sovranational/international entities and organizations	United Nations UNEP-FI United Nations Principles of Responsible Investment (UNPRI) United Nations Development Program (UNDP) OECD European Union People's Bank of China French Ministry of the Ecology	United Nations Global Compact Principles Partnership for Action in Green Economy UNPRI Private Equity Action on Climate Change UNDP- SDG Impact Practice Standards for Private Equity Funds UNDP-OECD—Impact Standards for Financing Sustainable Development OECD—Framework for SDG Aligned Finance EU Sustainable Finance Taxonomy Chinese Green Bond Catalogue Greenfin Label—France
Public and private finance institutions/organizations	Asian Development Bank (ADB) International Finance Corporation (IFC) Climate Bonds Initiatives (CBI) Green Climate Fund (GCF) Multilateral Development Banks (MDB) World Bank European Investment Bank	ADB Safeguard Policy Statement CBI standards and certification scheme GCF Investment Framework MDB Common Principles for Climate Mitigation World Bank Environmental and Social Framework Environmental and Social Standards

Typology	Name of issuer (main list)	Standard/framework outputs (main list)
Non-financial and non-governmental institutions	Global Reporting Initiative (GRI)	GRI Standards
	Loan Market Association (LMA)	LMA Green Loan Principles
	Equator Principles	Equator Principles (I, II, III, and IV)
	International Integrated Reporting Council (IIRC)	IIRC Integrated Reporting framework
	Financial Accounting Standards Board (FASB)	FASB Non Financial Reporting Standards
	International Sustainability Standards Board (ISSB)	SASB Investment and Commercial Banking Standard
	Sustainability Accounting Standard Board (SASB)	CiCERO Shades of Green
	CiCERO	TCFD – Climate Related Financial Disclosures
	Task Force for Climate Related Financial Disclosure (TCFD)	EFRAG Sustainability Reporting Standards Roadmap
	European Financial Reporting Advisory Group (EFRAG)	
Associations, Networks and think thank	Global Investor Coalition	The Low Carbon Registry
	Partnership for Carbon Accounting Financials (PCAF)	PCAF—Global GHG Accounting & Reporting Standard for the Financial Industry
	International Capital Market Association (ICMA)	ICMA Green Bond Principles; ICMA Sustainability Bond Guidelines
	Global Impact Investing Network	ICMA Harmonized Framework for Impact Reporting
	World Business Council For Sustainable Development (WBCSD)	WBCSD Framework for portfolio sustainability assessments
	ASEAN Capital Market Forum	ASEAN Green Bond Standards

Source Author's elaboration

All these standards deal with different aspects of the green sector, such as financial activities, products and services, investment phases, or risk management activities. They have been elaborated to provide a common language, frameworks, or procedures, by reducing transaction costs while also encouraging economic exchange. Specifically, five main themes have been identified (Gilchrist et al., 2021; Migliorelli & Marini, 2020; Weber, 2018; Weber & ElAlfy, 2019) in the green standards and framework arena:

- capital allocation measures;
- risk management;
- fiduciary duty and responsibility;
- reporting and disclosure;
- roadmaps for greening the financial system.

The first group concerns measures to sustain the capital allocation in the green sector, and its market development. Standards concerning risks aim to encourage environmental risk management practices within financial institutions, or in supervisory frameworks. Moreover, an emerging set of standards relates to identifying financial institutions' responsibilities regarding environmental factors within capital markets, as well as the relevance of environmental alignment issues in the context of financial institutions' fiduciary duties. The area of standards related to reporting and disclosure is the largest and most populated and frequently updated among green finance sectors; these are measures to disclosure flows of information relating to environmental factors within the financial system, such as the percentage of environment-aligned assets to total investment portfolios (Troeger & Steuer, 2021). Finally, initiatives related to the greening of the financial system are elaborated in roadmaps or plans, which aim to align the financial system with environmental sustainability objectives.

A recent clustering of green finance standards (Nedopil et al., 2021) provides a segmentation, according to:

- input-based standards (which are all the forms of green taxonomical efforts useful for awarding products a green label);
- process-based standards (mainly related to safeguards and performance measure activities);

- output-based standards (referring to the themes of emission thresholds and related activities of disclosure, reporting, and measurement of environmental impact).

In addition, among the emergent themes within the green standardization measures, three main topics have received growing attention: (*i*) taxonomy and related question of the portfolio's environmental alignment (House et al., 2020), (*ii*) the European sustainable finance framework (Wolf et al., 2021); and, (*iii*) the *greenwashing* issues (Gatti et al., 2021; Khan et al., 2020). In particular, many scholars (Gabor, 2021; Gimeno & Sols, 2020; Gortsos, 2021; Piebalgs & Jones, 2021) have explored both the theoretical and practical implications of adopting a green activities taxonomy in the asset allocation activities. Furthermore, the theme of *greenwashing* is analyzed in terms of limitations of the green finance market that may divert investors from this arena (Yu et al., 2020; Zeidan, 2020).

Finally, an important issue about setting innovative regulations and standards in green finance has emerged from the European Union context. Since the European Action plan for Sustainable Finance was issued to drive more long-term investment in sustainable economic activities, environmental issues have been in the spotlight, generally regarding such actions' implications for financial markets. Specifically, in July 2020 the EU introduced the Taxonomy Regulation, and the Sustainable Finance Disclosure Regulation on March 2021. The Taxonomy Regulation introduced a list of environmental economic activities, while the Disclosure Regulation imposed a mandatory disclosure and transparency framework for European financial actors. Furthermore, as part of the European Green Deal announced in December 2019, in April 2021, the European Commission adopted a package[6] of measures to make the European market a world-leading standard-setter for sustainable finance. The package includes:

---

[6] The Sustainable Finance package is an ambitious and comprehensive package of measures to help improve the flow of capitals to sustainable activities across the European Union. For further details: https://ec.europa.eu/info/publications/210421-sustainable-finance-communication_en.

- The EU Taxonomy Climate Delegated Act, which introduced, among others, mandatory disclosure for companies and financial actors;
- A proposal for a Corporate Sustainability Reporting Directive, which will extend the EU's sustainability reporting requirements to all listed companies;
- Six amending Delegated Acts.

Remaining in the European context, it is important to note that after the so-called "Brexit," the UK Government adopted an independent green finance regulation by launching the Green Finance Strategy in 2019 (UK Gov, 2019). Furthermore, in November 2021, the UK Government announced the adoption of mandatory disclosure of climate-related financial activities by 2025 (UK Gov, 2021).

2.6 The Green Finance Market: Size and Segments

The green finance market as a whole reflects the complexity and variety of its actors, instruments, and segments. This growing and uncoordinated arena makes it difficult to capture the entire size of green financial investments, which may vary according to the instruments used, the actors, and the investment focus. For these reasons, it is possible to provide a complete market overview by focusing on either (i) the main forms of green finance (i.e., green bonds, green lending, green equity investments), or (ii) mapping the segmentation of the green finance market sectors.

As regards green bonds, over the last 14 years, the majority of the bonds labeled as "green" are in line with the criteria outlined in the Green Bond Principles (GBP); most of them also receive verification from external reviewers. However, to certify the greenness of bonds, certifications and standard schemes are emerging, such as the Climate Bonds Standard and Certification Scheme, developed by the Climate Bonds Initiative (CBI). More recently, green bonds have been differentiated by the development of products defined with the following expressions: Sustainability, Transition, and Sustainability-linked bonds (Kölbel & Lambillon, 2022; Mocanu et al., 2021; Oxford Analytica,

Table 2.2 Green and sustainable bonds: an overview

Debt structure	Bond focus	Market standards/guidelines	Issuer's commitment
Use of Proceeds	Green	EU Green Bond Standards	Allocation, reporting
		ICMA Green Bond Principles	Allocation, reporting
	Sustainability	ICMA Sustainability Bond Guidelines	Allocation, reporting
Use of Proceeds OR General Corporate Sustainability's Purposes	SDGs	UNDP—SDG Impact Standards for Bond Issuers	Allocation, target achievement, and reporting
	Transition	Unformal (Contained in ICMA Transition Finance Handbook)	Allocation, target achievement, and reporting
General Corporate Sustainability's Purposes	Company's Sustainability Strategy	ICMA Sustainability-Linked Bond Principles	Allocation, target achievement, and reporting
	KPI—Linked Strategy	ICMA Sustainability-Linked Bond Principles	Allocation, reporting

Source Author's Elaboration

2021). According to Climate Bonds Market Intelligence,[7] the sustainable bond market, including all such types of bonds, amounted to just over half a trillion (USD 517.4bn) in 2021. However, conventional green bonds accounted for half of the sector's issuance up to October 2021 (in excess of USD 370bn), meaning a 49% growth rate in the five years before. Europe remains the hottest market for green and sustainable bond issuance, with over 50% of total issuance in the first half of 2021. In simplified terms, such populated arena of green and sustainable bonds cab be distinguished by their structure, as synthesized in Table 2.2.

The United States and Germany maintained their leading first and second positions among top issuing nations, followed by France and China. Of the companies issuing green and sustainability bonds in 2021, utilities dominated, followed by materials and consumer staples. The

---

[7] Available at: https://www.climatebonds.net/market-intelligence (accessed February 1, 2022).

Renewable Energy category drew the largest share of green investment across sectors and issuer types in 2021, followed by investments in Low-Carbon Buildings and Transport. Finally, JP Morgan, Goldman Sachs, and the Bank of America were the top-ranked banks for helping companies in green bond issuing (Standard & Poors, 2021).

Furthermore, for green loans, a structure evolution is represented by sustainability-linked loans. They are a relatively recent innovation, but volumes have risen dramatically over the past few years: to over USD 400 billion in 2021, from USD 220 billion in 2020, according to World Bank (2021). The growth of this kind of sustainable private debt derives from a combination of trends. Rising environmental and social pressure and market regulation have stimulated companies to issue these types of loans. A brief segmentation of green loans is reported in Table 2.3.

The European market is the most active for green loan issuing, representing about half of the market since 2015. Together with Asia, which is the second-largest market, they formed 80% of the green loan market in 2020.

Table 2.3 A segmentation of green loans

Classification based on debt structure	*Classification of green loans within banking sector*	*Classification of green loans based on specific green targets*
Use of Proceeds Green Loans Sustainability-Linked Loans	*Green Bilateral Loan* (formalized between corporate and bank) *Green syndicated loan* (a group of banks finance an operation, with one of them acting as green agent) *Green revolving credit facility* (line of credit based on ESG criteria which may influence the level of interest rate applied) *Green project finance* (based on long-term cash flows generated by a project or portfolio of projects identified by green eligible criteria)	Solar Power Loans Energy Efficiency Loans Waste Management Loans Reforestation Loans

Source Author's elaboration

Finally, regarding green equity, according to the Global Sustainable Investment Alliance (2020), around one-third of global equity investments (USD35.3 trillion, out of around USD 105 trillion in 2020) is made in sustainable green activities; mostly via index investing or equity funds. However, the true scale of green equity investments depends very much on the definition of "green" activities. While green indexes are transparent and standardized about the procedures for identifying green companies, this is not the case for green equity funds, which adopt heterogeneous and partially undisclosed identification strategies for their green eligible targets; especially in the green-listed equity segment (for instance: Shrimali, 2019). A further difficulty in determining the size of the green equity market is due to the diverse approaches adopted in the fund investment strategies, ranging from screening methods (positive or negative) to ESG integration (Table 2.4).

A further approach to capturing the green finance market's size is to explore the variety of the green finance sector segments. The sectors are heterogeneous and present different levels of maturity, and they largely correspond to the natural partition of green activities, with some exceptions concerning transverse sectors.

To provide an intuitive visualization, a map of green finance sector has been elaborated (Fig. 2.5) by grouping green finance sectors according to the four classical elements of nature (fire, air, water, and earth), as mentioned in Sect. 2.2.

The segment of biodiversity finance shows the fastest growth among non-mature sectors. According to OECD (2020), global biodiversity finance was estimated at USD 85 billion on average per year between 2015 and 2017. This estimate comprises public sector investments, international public expenditure, and private finance on biodiversity, estimated at USD 6.6–13.6 billion per year.

The clean energy finance segment represents the green finance sector with a mature expansion. According to the International Renewable Energy Agency (IRENA) (2020), renewable energy annual investment would have to almost triple (compared to 2020 data) to USD 800 billion by 2050. The private sector remains the main provider of capital for renewables, accounting for more than 85%; this sector is closely related to climate targets (adaptation and mitigation). According to International Development Finance Club (IFC) (2020), finance for green energy and mitigation of greenhouse gases was the largest category, representing 87%

Table 2.4 Main investment approaches in green equity markets

Investment approach	Description	Examples
Exclusionary/Negative Screening	This approach involves the screening of investments based on the exclusion of companies, sectors, or countries from the investible universe. This approach is also referred to as ethical-based exclusions	Exclusion criteria of companies involved in the production of weapons, tobacco, or in environmental crimes
Best-in-class	This approach involves the selection of best-performing companies of a specific investment universe	The selection weights companies based on ESG indexes or, e.g., on sustainability performance indexes
Thematic Selection	This approach addresses sustainability-linked assets or business sectors	Sustainability-themed Investments may be focused on a single or multiple issues related to ESG such as climate change, biodiversity or renewables
ESG Integration	This approach includes explicit consideration of ESG factors alongside financial analysis based on appropriate ESG data sources	The ESG funds focused on the potential impact of companies' ESG factors such as the efficient use of energy or the use of renewable energies that emit fewer GHG or the responsible management of waste
Environmental Impact Investing	This approach seeks the contemporary achievement of financial returns alongside a positive and measurable environmental impact	Environmental impact investing may include community energy investments, green buildings investments, or climate change impact funds

Source Author's elaboration

of climate finance among national and regional development banks' green investments.

The United Nations Environment Programme (UNEP) Finance Initiative (UNEP, 2020) defines circular finance as capitals that are exclusively used to finance, re-finance, invest in, or insure companies or projects that advance the circularity of the economy. According to Schröder and Raes

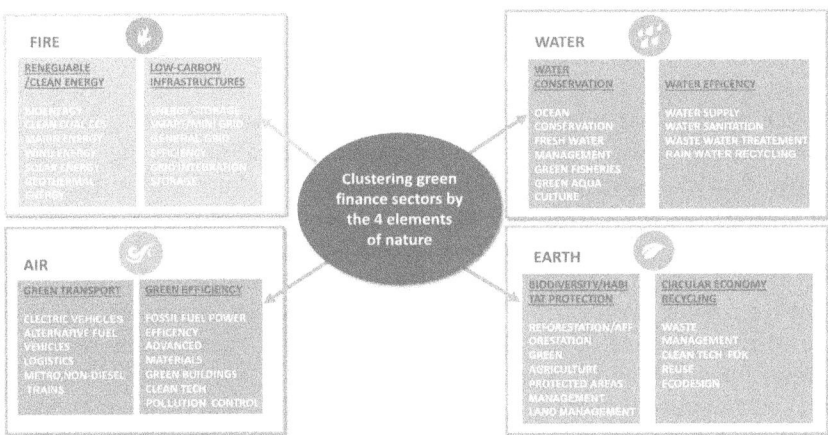

Fig. 2.5 A visualization of the green finance sectors through the four elements of nature (*Source* Author's elaboration)

(2021), governments, companies, and private financial institutions invest a yearly total of USD 1.3trn in circular economy initiatives, to reduce waste and promote reuse and recycling; for instance, USD 800bn annually is funded by corporate finance, and USD 46bn from financial institutions.

Finally, regarding the "water" quadrant of the map, the blue and ocean finance is increasingly attracting investors, insurers, banks, and policymakers, as a new source of prosperity. Its annual economic value is estimated at USD 2.5 trillion (WHO, 2017), equivalent to the world's seventh-largest economy, and it is beginning to drive a green finance sector in full growth, whose market is estimated at USD3 trillion (Cisneros-Montemayor et al., 2019; Mohanty, 2018).

2.7 Conclusions

This chapter introduced the topic of green finance by summarizing the state of the art of this newly formed interdisciplinary field by setting out definitional boundaries and a conceptual framework, including the instruments, markets, and actors in this field. The market ecosystem is analyzed based on the results of the literature review. Specifically, a segmentation of the market is derived and recent regulation is focused upon. In addition, the chapter outlined asset classes, including the green finance spans and typologies of the financial instruments employed.

From the analysis of the results emerged how the definition of green finance has evolved over time, shifting from macro considerations related to a vision of finance that promotes environmentalism, to the greening of the financial system to tackle environmental issues. After the Paris Agreement in 2015, research in green finance has entered a vigorous stage, and the focus of green finance definitions has started to focus in the integration of environmental factors into investment decisions, for long-term value creation. However, a certain level of ambiguity remains, due to the existence of overlapping terms. Indeed, "sustainable finance," "green finance," and "climate finance" are often used interchangeably, and are therefore misunderstood. Moreover, various stakeholders to identify green finance sometimes use other terms such as "ESG investments and responsible investments."

Furthermore, the definitional puzzle of green finance is related to a second dimension: the identification of what "green" signifies in the context of green finance. This element represents a crucial step for the advancement of the field because all the frameworks, taxonomies, and eligibility criteria adopted for green ratings and the use of proceeds are based on different forms of intended green activities. A wider agreed definition of green finance is therefore central to the development of green finance as an academic research field, as well as a market segment within the broader sector of finance for sustainable development.

References

Adhami, S., Giudici, G., & Anh, H. P. (2017, November). *Crowdfunding for green projects in Europe: Success factors and effects on the local environmental performance and wellbeing* (Rep.). Retrieved April 25, 2019, from http://www.crowdfundres.eu/wp-content/uploads/2017/11/Crowdfunding-for-green-projects-in-Europe-2017.pdf

Ahlström, H., & Monciardini, D. (2021). The regulatory dynamics of sustainable finance: Paradoxical success and limitations of EU reforms. *Journal of Business Ethics*, 1–20.

Akomea-Frimpong, I., Adeabah, D., Ofosu, D., & Tenakwah, E. J. (2021). A review of studies on green finance of banks, research gaps and future directions. *Journal of Sustainable Finance & Investment*, 1–24.

Albino, V., Balice, A., & Dangelico, R. M. (2009). Environmental strategies and green product development: An overview on sustainability-driven companies. *Business Strategy and the Environment, 18*(2), 83–96.

Anderson, J. (2016). Environmental finance. In *Handbook of environmental and sustainable finance* (pp. 307–333). Academic Press.

Babiak, K., & Trendafilova, S. (2011). CSR and environmental responsibility: Motives and pressures to adopt green management practices. *Corporate Social Responsibility and Environmental Management, 18*(1), 11–24.

Banga, J. (2019). The green bond market: A potential source of climate finance for developing countries. *Journal of Sustainable Finance & Investment, 9*(1), 17–32.

Berensmann, K., & Lindenberg, N. (2019). Green finance: Across the universe. In *Corporate social responsibility, ethics and sustainable prosperity* (pp. 305–332). World Scientific Publishing.

Berger, R. (2011). *Green growth, green profit: How green transformation boosts business*. Palgrave Macmillan UK.

Berrou, R., Ciampoli, N., & Marini, V. (2019). Defining green finance: Existing standards and main challenges. In *The rise of green finance in Europe* (pp. 31–51). Palgrave Macmillan.

Biesbroek, G. R., Swart, R. J., & Van der Knaap, W. G. (2009). The mitigation–adaptation dichotomy and the role of spatial planning. *Habitat International, 33*(3), 230–237.

Boffo, R., & Patalano, R. (2020). *ESG investing: Practices, progress and challenges*. OECD. www.oecd.org/finance/ESG-Investing-Practices-Progress-and-Challenges.pdf

Bongardt, A., & Torres, F. (2022). The European Green Deal: More than an exit strategy to the pandemic crisis, a building block of a sustainable European economic model. *JCMS: Journal of Common Market Studies, 60*(1), 170–185.

Bonsu, N. O. (2020). Towards a circular and low-carbon economy: Insights from the transitioning to electric vehicles and net zero economy. *Journal of Cleaner Production, 256*, 120659.

Bos, M., Pressey, R. L., & Stoeckl, N. (2015). Marine conservation finance: The need for and scope of an emerging field. *Ocean & Coastal Management, 114*, 116–128.

Bracking, S. (2015). The anti-politics of climate finance: The creation and performativity of the green climate fund. *Antipode, 47*(2), 281–302.

Bridge, G., Bouzarovski, S., Bradshaw, M., & Eyre, N. (2013). Geographies of energy transition: Space, place and the low-carbon economy. *Energy Policy, 53*, 331–340.

Brühl, V. (2021). *Green finance in Europe—Strategy, regulation and instruments* (Center for Financial Studies Working Paper, 657).

Bryant, G. (2019). *Carbon markets in a climate-changing capitalism*. Cambridge University Press.
Buchner, B., Falconer, A., Hervé-Mignucci, M., Trabacchi, C., & Brinkman, M. (2011, October 27). *The landscape of climate finance*. Climate Policy Initiative.
Chang, C. E., Nelson, W. A., & Witte, H. D. (2012). Do green mutual funds perform well? *Management Research Review, 35*(8), 693–708.
Chiappini, H. (2017). An introduction to social impact investing. In *Social impact funds*. Palgrave Studies in Impact Finance. Palgrave Macmillan. https://doi.org/10.1007/978-3-319-55260-6_2
Cho, Y. N., & Taylor, C. R. (2020). The role of ambiguity and skepticism in the effectiveness of sustainability labeling. *Journal of Business Research, 120*, 379–388.
Cisneros-Montemayor, A. M., Moreno-Báez, M., Voyer, M., Allison, E. H., Cheung, W. W., Hessing-Lewis, M., Oyinlola, M. A., Singh, G. G., Swartz, W., & Ota, Y. (2019). Social equity and benefits as the nexus of a transformative Blue Economy: A sectoral review of implications. *Marine Policy, 109*, 103702.
Clapp, J., & Stephens, P. (2019). Financialising nature. In *Routledge handbook of global sustainability governance* (pp. 206–217). Routledge.
Clark, R., Reed, J., & Sunderland, T. (2018). Bridging funding gaps for climate and sustainable development: Pitfalls, progress and potential of private finance. *Land Use Policy, 71*, 335–346.
Clark, S. (2012). *A field guide to conservation finance*. Island Press.
Collinson, P. (2015). Fossil fuel-free funds outperformed conventional ones, analysis shows. *The Guardian*, 10.
Cui, Y., Geobey, S., Weber, O., & Lin, H. (2018). The impact of green lending on credit risk in China. *Sustainability, 10*(6), 2008.
Cumming, D. J., Leboeuf, G., & Schwienbacher, A. (2017). Crowdfunding cleantech. *Energy Economics, 65*, 292–303.
De la Porte, C., & Jensen, M. D. (2021). The next generation EU: An analysis of the dimensions of conflict behind the deal. *Social Policy & Administration, 55*(2), 388–402.
Degryse, H., Goncharenko, R., Theunisz, C., & Vadasz, T. (2021). *When green meets green*. Available at SSRN 3724237.
Devas, H. (1994). Green Finance. *European Energy and Environmental Law Review, 3*, 220.
Dietz, S., Bowen, A., Dixon, C., & Gradwell, P. (2016). 'Climate value at risk' of global financial assets. *Nature Climate Change, 6*(7), 676–679.
D'Orazio, P., & Popoyan, L. (2019). Fostering green investments and tackling climate-related financial risks: Which role for macroprudential policies? *Ecological Economics, 160*, 25–37.

Doumbia, D., & Lauridsen, M. L. (2019). Closing the SDG financing gap: Trends and data. *EMCompass, 73*, 1–8.

Eckert, E., & Kovalevska, O. (2021). Sustainability in the European Union: Analyzing the discourse of the European Green Deal. *Journal of Risk and Financial Management, 14*(2), 80.

Ekins, P. (2002). *Economic growth and environmental sustainability: The prospects for green growth*. Routledge.

European Banking Federation (EBF). (2017). *Towards a green finance framework*. https://www.ebf.eu/wp-content/uploads/2017/09/Geen-finance-complete.pdf

European Union Platform on Sustainable Finance. (2021). *Draft Report by Subgroup 4: Social Taxonomy*. https://ec.europa.eu/info/sites/default/files/business_economy_euro/banking_and_finance/documents/sf-draft-report-social-taxonomy-july2021_en.pdf

Falcone, P. M., Morone, P., & Sica, E. (2018). Greening of the financial system and fuelling a sustainability transition: A discursive approach to assess landscape pressures on the Italian financial system. *Technological Forecasting and Social Change, 127*, 23–37.

Fischer, R. (2014). *Demystifying Private Climate Finance*. UNEP. https://stg-wedocs.unep.org/bitstream/handle/20.500.11822/9339/-Demystifying_Private_Climate_.pdf?sequence=2

Fuchs, S., Kachi, A., Sidner, L., & Westphal, M. (2021). *Aligning financial intermediary investments with the Paris Agreement*. World Resources Institute.

Fu, J., & Ng, A. W. (2020). Green finance reform and innovation for sustainable development of the Greater Bay Area: Towards an ecosystem for sustainability. In *Sustainable energy and green finance for a low-carbon economy* (pp. 3–23). Springer.

Gabor, D. (2021). The wall street consensus. *Development and Change, 52*(3), 429–459.

Galanakis, C. M., Brunori, G., Chiaramonti, D., Matthews, R., Panoutsou, C., & Fritsche, U. R. (2022). Bioeconomy and green recovery in a post-COVID-19 era. *Science of the Total Environment, 808*, 152180.

Gatti, L., Pizzetti, M., & Seele, P. (2021). Green lies and their effect on intention to invest. *Journal of Business Research, 127*, 228–240.

Gibbs, D. (2003). Reconciling economic development and the environment. *Local Environment, 8*(1), 3–8.

Gilchrist, D., Yu, J., & Zhong, R. (2021). The limits of green finance: A survey of literature in the context of green bonds and green loans. *Sustainability, 13*(2), 478.

Gimeno, R., & Sols, F. (2020). Incorporating sustainability factors into asset management. *Financial Stability Review, 39*, 179–199.

Giraudet, L. G., Petronevich, A., & Faucheux, L. (2021). Differentiated green loans. *Energy Policy, 149*, 111861.

Global Sustainable Investment Alliance. (2020). *Global Sustainable Investment Review 2020*. http://www.gsi-alliance.org/wp-content/uploads/2021/08/GSIR-20201.pdf

Gond, J. P., & Piani, V. (2013). Enabling institutional investors' collective action: The role of the principles for responsible investment initiative. *Business & Society, 52*(1), 64–104.

Goovaerts, L., & Verbeek, A. (2018). Sustainable banking: Finance in the circular economy. In *Investing in resource efficiency* (pp. 191–209). Springer.

Gortsos, C. V. (2021). The Taxonomy Regulation: More important than just as an element of the capital markets union. In *Sustainable finance in Europe* (pp. 351–395). Palgrave Macmillan.

Green Finance Platform. (2021). *A New Era of Government Intervention on Green Finance Measures*. https://www.greenfinanceplatform.org/news/new-era-government-intervention-green-finance-measures

Gupta, J., & Vegelin, C. (2016). Sustainable development goals and inclusive development. *International Environmental Agreements: Politics, Law and Economics, 16*(3), 433–448.

G20 Green Finance Study Group. (2016). *G20 Green Finance Synthesis Report*. https://g20sfwg.org/wp-content/uploads/2021/07/2016_Synthesis_Report_Full_EN.pdf

Hafner, S., James, O., & Jones, A. (2019). A scoping review of barriers to investment in climate change solutions. *Sustainability, 11*(11), 3201.

Hafner, S., Jones, A., Anger-Kraavi, A., & Pohl, J. (2020). Closing the green finance gap—A systems perspective. *Environmental Innovation and Societal Transitions, 34*, 26–60.

Haigh, N., & Hoffman, A. J. (2011). Hybrid organizations: The next chapter in sustainable business. *Organizational Dynamics, 41*(2), 126–134.

Heaps, T., & Guyatt, D. (2017). *A review of international financial standards as they relate to sustainable development*. UN Environment Inquiry.

He, L., Liu, R., Zhong, Z., Wang, D., & Xia, Y. (2019). Can green financial development promote renewable energy investment efficiency? A consideration of bank credit. *Renewable Energy, 143*, 974–984.

Herbertson, K., & Hunter, D. (2006). Emerging standards for sustainable finance of the energy sector. *Sustainable Development Law and Policy, 7*, 4.

Höhne, N., Khosla, S., Fekete, H., & Gilbert, A. (2012). *Mapping of Green Finance Delivered by IDFC Members in 2011*. https://www.idfc.org/wp-content/uploads/2019/03/idfc_green_finance_mapping_report_2012_06-14-12.pdf

House, K., Mok, L. W., & Tripathy, A. (2020). Defining climate-aligned investment: An analysis of sustainable finance taxonomy development. *Journal of Environmental Investing, 10*(1), 80.

Hutton, G., & Varughese, M. (2016). *The costs of meeting the 2030 sustainable development goal targets on drinking water, sanitation, and hygiene*. https://openknowledge.worldbank.org/bitstream/handle/10986/23681/K8543.pdf?sequence=1

Illman, J., Halonen, M., Rinne, P., Huq, S., & Tveitdal, S. (2013). *Scoping study on financing adaptation-mitigation synergy activities*.

Inderst, G., Kaminker, C., & Stewart, F. (2012). *Defining and measuring green investments: Implications for institutional investors' asset allocations* (OECD Working Papers on Finance, Insurance and Private Pensions, No. 24). OECD Publishing.

International Capital Market Association (ICMA). (2021). *Green Bond Principles*. Voluntary Process Guidelines for Issuing Green Bonds. https://www.icmagroup.org/sustainable-finance/the-principles-guidelines-and-handbooks/green-bond-principles-gbp/

International Development Finance Club (IFC). (2020). *IDFC Green Finance Mapping Report 2020*. https://www.idfc.org/wp-content/uploads/2020/11/idfc-2020-gfm-full-report_final-1.pdf

Jaumotte, M. F., Liu, W., & McKibbin, W. J. (2021). *Mitigating climate change: Growth-friendly policies to achieve net zero emissions by 2050* (No. 16553). International Monetary Fund.

Jones, A. W. (2015). Perceived barriers and policy solutions in clean energy infrastructure investment. *Journal of Cleaner Production, 104*, 297–304.

Jordan, A., Huitema, D., Van Asselt, H., Rayner, T., & Berkhout, F. (Eds.). (2010). *Climate change policy in the European Union: Confronting the dilemmas of mitigation and adaptation?*. Cambridge University Press.

Kaneko, J. (2021). Ethics in education for sustainable finance: Challenges toward long-termism in Japan and Europe. In *Handbook on Ethics in Finance* (pp. 247–268). Springer.

Kay, K. (2018). A hostile takeover of nature? Placing value in conservation finance *Antipode, 50*(1), 164–183.

Khan, H. Z., Bose, S., Mollik, A. T., & Harun, H. (2020). "Green washing" or "authentic effort"? An empirical investigation of the quality of sustainability reporting by banks. *Accounting, Auditing & Accountability Journal, 34*(2), 338–369.

Knight, E. R. (2011). Five perspectives on an emerging market: Challenges with clean tech private equity. *The Journal of Alternative Investments, 14*(3), 76–81.

Kölbel, J. F., & Lambillon, A. P. (2022, January 12). Who pays for sustainability? An analysis of sustainability-linked bonds. *An Analysis of Sustainability-Linked Bonds*.

Krämer, L. (2020). Planning for climate and the environment: The EU green deal. *Journal for European Environmental & Planning Law, 17*(3), 267–306.

Kublicki, N. M. (1993).Green finance: Problems and solutions concerning alternative environmental debt exchanges. *Vermont Law Review, 18*, 313.

Kurznack, L., Schoenmaker, D., & Schramade, W. (2021). A model of long-term value creation. *Journal of Sustainable Finance & Investment*, 1–19.

Labatt, S., & White, R. R. (2002). *Environmental finance: A guide to environmental risk assessment and financial products* (Vol. 98). Wiley.

Lagoarde-Segot, T. (2020). Financing the sustainable development goals. *Sustainability, 12*(7), 2775.

Lam, P. T., & Law, A. O. (2016). Crowdfunding for renewable and sustainable energy projects: An exploratory case study approach. *Renewable and Sustainable Energy Reviews, 60*, 11–20.

La Torre, M., & Chiappini, H. (2020). Contemporary issues in sustainable finance. In *Creating an efficient market through innovative policies and instruments*. Palgrave Macmillan.

Lehner, O. M. (2016). *Routledge handbook of social and sustainable finance*. Routledge.

Lesser, K., Lobe, S., & Walkshäusl, C. (2014). Green and socially responsible investing in international markets. *Journal of Asset Management, 15*(5), 317–331.

Liagre, L., Lara Almuedo, P., Besacier, C., & Conigliaro, M. (2015). *Sustainable financing for forest and landscape restoration: Opportunities, challenges and the way forward*. FAO/UNCCD.

Lindenberg, N. (2014). *Definition of green finance*.

Liu, Y. Y., & Keleher, T. (2009, November). *Green equity toolkit: Standards and strategies for advancing race, gender and economic equity in the green economy*. Gender and Economic Equity in the Green Economy.

Li, Z., Liao, G., Wang, Z., & Huang, Z. (2018). Green loan and subsidy for promoting clean production innovation. *Journal of Cleaner Production, 187*, 421–431.

Loan Market Association (LMA), Asia Pacific Loan Market Association (APLMA) and Loan Syndications & Trading Association (LSTA). (2021). *Green Loan Principles*. https://www.lma.eu.com/application/files/9716/1304/3740/Green_Loan_Principles_Feb2021_V04.pdf

Loiseau, E., Saikku, L., Antikainen, R., Droste, N., Hansjürgens, B., Pitkänen, K., Leskinen, P., Kuikman, P., & Thomsen, M. (2016). Green economy and related concepts: An overview. *Journal of Cleaner Production, 139*, 361–371.

Louche, C., Busch, T., Crifo, P., & Marcus, A. (2019). Financial markets and the transition to a low-carbon economy: Challenging the dominant logics. *Organization & Environment, 32*(1), 3–17.

Luzio, J. P. P., & Lemke, F. (2013). Exploring green consumers' product demands and consumption processes: The case of Portuguese green consumers. *European Business Review, 25*(3), 281–300.

Martin, M. (2015). Building impact businesses through hybrid financing. *Entrepreneurship Research Journal, 5*(2), 109–126.

Mazzucato, M., & Semieniuk, G. (2018). Financing renewable energy: Who is financing what and why it matters. *Technological Forecasting and Social Change, 127*, 8–22.

Mekonnen, A. (2014). Economic costs of climate change and climate finance with a focus on Africa. *Journal of African Economies, 23*(suppl_2), ii50–ii82.

Mendez, A., & Houghton, D. P. (2020). Sustainable banking: The role of multilateral development banks as norm entrepreneurs. *Sustainability, 12*(3), 972.

Meyer, C. A. (1997). Public-nonprofit partnerships and North-South green finance. *The Journal of Environment & Development, 6*(2), 123–146.

Meyers, D., Alliance, C. F., Bohorquez, J., Cumming, B. F. I. B., Emerton, L., Riva, M., Fund, U. J. S., & Victurine, R. (2020). Conservation finance: A framework. *Conservation Finance Alliance*, 1–45.

Migliorelli, M., & Marini, V. (2020). Sustainability-related risks, risk management frameworks and non-financial disclosure. In *Sustainability and Financial Risks* (pp. 93–118). Palgrave Macmillan.

Min, H., & Galle, W. P. (2001). Green purchasing practices of US firms. *International Journal of Operations & Production Management, 21*(9), 1222–1238.

Mocanu, M., Constantin, L. G., & Cernat-Gruici, B. (2021). Sustainability bonds. An international event study. *Journal of Business Economics and Management, 22*(6), 1551–1576.

Mohanty, S. K. (2018). Towards estimation of the blue economy. *The Blue Economy Handbook of the Indian Ocean Region, 64*.

Morseletto, P. (2020). Targets for a circular economy. *Resources, Conservation and Recycling, 153*, 104553.

Nakhooda, S., Watson, C., & Schalatek, L. (2015). *The global climate finance architecture*. Overseas Development Institute.

Nedopil, C., Dordi, T., & Weber, O. (2021). The nature of global green finance standards—Evolution, differences, and three models. *Sustainability, 13*(7), 3723.

Nigam, N., Mbarek, S., & Benetti, C. (2018). Crowdfunding to finance eco-innovation: Case studies from leading renewable energy platforms. *Journal of Innovation Economics Management, 2*, 195–219.

Ng, A. W. (2018). From sustainability accounting to a green financing system: Institutional legitimacy and market heterogeneity in a global financial centre. *Journal of Cleaner Production, 195*, 585–592.

Organization for Economic Cooperation and Development (OECD). (2017). *Investing in climate, investing in growth.* https://www.oecd.org/env/cc/g20-climate/synthesis-investing-in-climate-investing-in-growth.pdf

Organization for Economic Cooperation and Development (OECD). (2020). *A comprehensive overview of global biodiversity finance.* https://www.oecd.org/environment/resources/biodiversity/report-a-comprehensive-overview-of-global-biodiversity-finance.pdf

Organization for Economic Cooperation and Development (OECD) and Bloomber Philantropies. (2015). *Green investment banks. Policy Perspectives.* https://www.oecd.org/environment/cc/Green-Investment-Banks-POLICY-PERSPECTIVES-web.pdf

Oxford Analytica. (2021). Transition bonds market will grow from slow start. *Emerald Expert Briefings* (oxan-db).

Owen, R., Brennan, G., & Lyon, F. (2018). Enabling investment for the transition to a low carbon economy: Government policy to finance early stage green innovation. *Current Opinion in Environmental Sustainability, 31*, 137–145.

Park, H., & Kim, J. D. (2020). Transition towards green banking: Role of financial regulators and financial institutions. *Asian Journal of Sustainability and Social Responsibility, 5*(1), 1–25.

Petry, J., Fichtner, J., & Heemskerk, E. (2021). Steering capital: The growing private authority of index providers in the age of passive asset management. *Review of International Political Economy, 28*(1), 152–176.

Piebalgs, A., & Jones, C. (2021). *The importance of the EU taxonomy: The example of electricity storage.* European University Institute.

Pricewaterhouse Coopers Consultants (PWC). (2013). *Exploring Green Finance Incentives in China.* https://www.pwchk.com/en/migration/pdf/green-finance-incentives-oct2013-eng.pdf

Randjelovic, J., O'Rourke, A. R., & Orsato, R. J. (2003). The emergence of green venture capital. *Business Strategy and the Environment, 12*(4), 240–253.

Rezec, M., & Scholtens, B. (2017). Financing energy transformation: The role of renewable energy equity indices. *International Journal of Green Energy, 14*(4), 368–378.

Rhodes, C. J. (2016). The 2015 Paris climate change conference: COP21. *Science Progress, 99*(1), 97–104.

Sachs, J., Woo, W. T., Yoshino, N., & Taghizadeh-Hesary, F. (Eds.). (2019). *Handbook of green finance: Energy security and sustainable development.* Springer.

Schepers, D. H. (2011). The equator principles: A promise in progress? *Corporate Governance: The International Journal of Business in Society, 11*(1), 90–106.

Schmittmann, J., & Chua, H. T. (2021). *How green are green debt issuers?* International Monetary Fund.

Scholtens, B. (2011, August). The sustainability of green funds. In *Natural resources forum* (Vol. 35, No. 3, pp. 223–232). Blackwell Publishing.

Schröder, P., & Raes, J. (2021). *Financing an inclusive circular economy De-risking investments for circular business models and the SDGs* (Environment and Society Programme Research Paper). https://www.chathamhouse.org/sites/default/files/2021-07/2021-07-16-inclusive-circular-economy-schroder-raes.pdf

Schub, J. (2015). Green banks: Growing clean energy markets by leveraging private investment with public financing. *The Journal of Structured Finance, 21*(3), 26–35.

Shrimali, G. (2019). Do clean energy (equity) investments add value to a portfolio? *Green Finance, 1*(2), 188–204.

Sonntag-O'Brien, V., & Usher, E. (2012). Mobilizing finance for renewable energies. In *Renewable Energy* (pp. 197–223). Routledge.

Stahel, W. R. (2016). The circular economy. *Nature, 531*(7595), 435–438.

Standard & Poors. (2021). *Banks, firms attracted $175 bil by issuing climate-focused green bonds in 2021: IHS Markit analysts.* https://cleanenergynews.ihsmarkit.com/research-analysis/banks-firms-attracted-175-bil-by-issuing-climatefocused-green-.html

Sullivan, S. (2018). Making nature investable: From legibility to leverageability in fabricating 'nature' as 'natural capital'. *Science and Technology Studies, 31*(3), 47–76.

Sweet, W. (2016). *Climate diplomacy from Rio to Paris: The effort to contain global warming.* Yale University Press.

Taghizadeh-Hesary, F., & Yoshino, N. (2019). The way to induce private participation in green finance and investment. *Finance Research Letters, 31*, 98–103.

Tang, D. Y., & Zhang, Y. (2020). Do shareholders benefit from green bonds? *Journal of Corporate Finance, 61*, 101427.

Trapp, C. T., & Kanbach, D. K. (2021). Green entrepreneurship and business models: Deriving green technology business model archetypes. *Journal of Cleaner Production, 126694.*

Troeger, T. H., & Steuer, S. (2021). *The role of disclosure in green finance* (European Corporate Governance Institute-Law Working Paper, 604).

Umar, M., Ji, X., Mirza, N., & Naqvi, B. (2021). Carbon neutrality, bank lending, and credit risk: Evidence from the Eurozone. *Journal of Environmental Management, 296*, 113156.

UNEP. (1992). *Statement of Commitment by Financial Institutions (FI) on Sustainable Development.*

UNEP. (2020). *Financing circularity: Demystifying finance for the circular economy*. https://www.unepfi.org/publications/general-publications/financing-circularity/

United Nations. (2016). *Transforming our world: The 2030 agenda for sustainable development*. unepswiosm1inf7sdg.pdf

United Nations—Water Global analysis and assessment of sanitation and drinking water (UN GLAAS). (2017). *Financing universal water, sanitation and hygiene under the sustainable development goals* (GLAAS 2017 Report). https://apps.who.int/iris/bitstream/handle/10665/254999/9789241512190-eng.pdf

UK Government. (2019). *Green Finance Strategy. Trasforming finance for a greener future*. https://www.gov.uk/government/publications/green-finance-strategy

UK Government. (2021). *Greening Finance: A roadmap to sustainable investing*. https://assets.publishing.service.gov.uk/government/uploads/system/uploads/attachment_data/file/1031805/CCS0821102722-006_Green_Finance_Paper_2021_v6_Web_Accessible.pdf

Van Duuren, E., Plantinga, A., & Scholtens, B. (2016). ESG integration and the investment management process: Fundamental investing reinvented. *Journal of Business Ethics, 138*(3), 525–533.

Van Steenis, H. (2019). Defective data is a big problem for sustainable investing. *Financial Times website*. https://www.ft.com/content/c742edfa-30be-328e-8bd2-a7f8870171e4

Vickers, I., & Lyon, F. (2014). Beyond green niches? Growth strategies of environmentally-motivated social enterprises. *International Small Business Journal, 32*(4), 449–470.

Xepapadeas, A. (2005). Economic growth and the environment. *Handbook of Environmental Economics, 3*, 1219–1271.

Yuan, F., & Gallagher, K. P. (2018). Greening development lending in the Americas: Trends and determinants. *Ecological Economics, 154*, 189–200.

Yuan, Y., Cai, F., & Yang, L. (2020). Renewable energy investment under carbon emission regulations. *Sustainability, 12*(17), 6879.

Yu, E. P. Y., Van Luu, B., & Chen, C. H. (2020). Greenwashing in environmental, social and governance disclosures. *Research in International Business and Finance, 52*, 101192.

Walley, E. E., & Taylor, D. W. (2002). Opportunists, champions, mavericks…? A typology of green entrepreneurs. *Greener Management International, 38*, 31–43.

Wang, C., Nie, P. Y., Peng, D. H., & Li, Z. H. (2017). Green insurance subsidy for promoting clean production innovation. *Journal of Cleaner Production, 148*, 111–117.

Wang, Y., & Zhi, Q. (2016). The role of green finance in environmental protection: Two aspects of market mechanism and policies. *Energy Procedia, 104*, 311–316.

Waters, R. (2018). *Clean tech unicorn Bloom Energy limps on to stock market.* https://www.ft.com/content/18562814-9017-11e8-b639-7680cedcc421

Weber, O. (2018). Financial sector sustainability regulations and voluntary codes of conduct: Do they help to create a more sustainable financial system? In *Designing a sustainable financial system* (pp. 383–404). Palgrave Macmillan.

Weber, O., & ElAlfy, A. (2019). The development of green finance by sector. In M. Migliorelli & P. Dessertine (Eds.), *The rise of green finance in Europe*. Palgrave Studies in Impact Finance. Palgrave Macmillan. https://doi.org/10.1007/978-3-030-22510-0_3

Weber, O., & Remer, S. (2011). *Social banks and the future of sustainable finance*. Routledge.

Wolf, S., Teitge, J., Mielke, J., Schütze, F., & Jaeger, C. (2021). The European Green Deal—More than climate neutrality. *Intereconomics, 56*(2), 99–107.

World Bank. (2017). *Climate Finance Overview*. Report. http://www.worldbank.org/en/topic/climatefinance

World Bank. (2021). *What you need to know about green loans*. https://www.worldbank.org/en/news/feature/2021/10/04/what-you-need-to-know-about-green-loans

World Health Organization. (2017). *Radical Increase in Water and Sanitation Investment Required to Meet Development Targets*. https://www.who.int/news/item/13-04-2017-radical-increase-in-water-and-sanitation-investment-required-to-meet-development-targets#:~:text=Radical%20increase%20in%20water%20and%20sanitation%20investment%20required%20to%20meet%20development%20targets,-13%20April%202017&text=In%20order%20to%20meet%20the,include%20operating%20and%20maintenance%20costs.

Zadek, S., & Flynn, C. (2013). *South-originating green finance: Exploring the potential*. https://doc.rero.ch/record/208970/files/06-south-originating.pdf

Zeidan, R. (2020). Obstacles to sustainable finance and the covid19 crisis. *Journal of Sustainable Finance & Investment, 12*(2), 525–528.

Zhou, X. Y., Caldecott, B., Hoepner, A. G., & Wang, Y. (2020). Bank green lending and credit risk: An empirical analysis of China's Green Credit Policy. *Business Strategy and the Environment, 31*(4), 1623–1640.

Ziolo, M., Filipiak, B. Z., Bąk, I., & Cheba, K. (2019). How to design more sustainable financial systems: The roles of environmental, social, and governance factors in the decision-making process. *Sustainability, 11*(20), 5604.

CHAPTER 3

The Green Financing Framework Combining Innovation and Resilience: A Growing Toolbox of Green Finance Instruments

3.1 Introduction

The green economy has enormous potential to drive sustainable growth and job creation and, especially after the spread of COVID-19 and the related green-targeted economic responses, to increase the sector's traction in the business world (Enoch, 2021; Erdelen & Richardson, 2021). Many businesses in different sectors have already started to tap into this potential, displaying a wide variety of innovative green technologies and green-based business models (Chevallier et al., 2021; Trapp & Kanbach, 2021). However, the transition to a net-zero economy requires tackling a number of hurdles; these include ensuring access to finance for innovative green businesses, throughout their various growth phases (Stern & Stiglitz, 2022).

Transitioning to a green-based economy requires nothing less than an economic paradigm shift from an economic development model that is unsustainable, both economically and environmentally, to a more long-term value approach to investments, where different forms of non-financial returns are assessed (Bhattacharyya, 2022). Green finance attempts to encompass all economic systems where the resources used for a product or a service are environmentally friendly, in terms of delivering environmental benefits in the short, medium, and long term (Sinha et al., 2021). Green finance-related products focus on the financing of green products, processes, value chains, business, and service models, in order to achieve the above-specified purpose (Huang et al., 2022).

It is estimated that transitioning to a low-carbon economy in the next decade will require significant investment to be delivered by private investors on a much larger scale than previously, considering the limitations in public budgeting (Shideler & Hetzel, 2021). Green projects are generally eligible for green financing, as these are well aligned with one or more of the regulatory or policy objectives. Indeed, there are a number of different operational definitions of green investments provided for different regulatory, geographical, and government activities. Thus, the chapter provides an overview of the existing green financing framework. In addition, some exemplar case studies of the most innovative green financing emerging trends are illustrated.

Accordingly, this chapter has been organized as follows. Section 3.2 presents an overview of the green financing framework, while Sect. 3.3 focuses on green financing delivery instruments. Section 3.4 presents the innovative tools and markets emerging within the green financing landscape, and Sect. 3.5 presents illustrative case studies of these still unexplored tools, by highlighting practical implications of the green finance market's development. Finally, Sect. 3.6 concludes.

3.2 The Green Financing Framework

Mapping the current green financial flows from the financial sector, and relating them to the existing demand for green finance, depends on the financial instrument under consideration, as well as on how one defines what activities are "green" (Ba, 2022). Furthermore, to help those who supply green capitals to identify their green targets, the financial intermediation chain includes actors such as green rating providers, who provide "green" scores of equity and debt issuers based on their disclosures; and index providers, who convert ratings into market indices by reweighting market portfolios according to the investor's investment approaches (Neisen et al., 2021). There are a number of crucial financial intermediaries and institutions driving the green finance market, including banks, institutional investors, and international financial institutions, which can be both private and public entities (Migliorelli & Dessertine, 2019).

Banking system assets play an important role within the green financial market, representing an important share of global green financial assets

(Del Gaudio et al., 2022; Raberto et al., 2019). Within this context, it is important to distinguish two main broad categories of the banking sector's presence in the green finance market. These banks based on a green/sustainable financial business model, and those deliver a limited percentage of green capital from their total financial assets. In the first case, it is possible to include banks identified as "social," "ethical," "cooperative," or "alternative" (Gangi et al., 2021; Höhnke & Homölle, 2021; Montes, 1998; Tan et al., 2017; Weber, 2014). Such banks are based on business models for sustainability (Singh & Jayaram, 2021; Yip & Bocken, 2018), which focus not on maximizing profits but on maximizing value for a wide audience of stakeholders (Győri et al., 2021; Kocornik-Mina et al., 2021). On the other hand, one can find traditional or commercial banks which direct a percentage of their capitals toward green activities. Such banks are based on business models designed to maximize the value for their shareholders.

More recently, a grey zone between the two sides has emerged, identified with the expression "green banking," to indicate banks that are concerned about environmental well-being while financing businesses (Biswas, 2011; Bukhari et al., 2022; Gupta, 2015; Sharma & Choubey, 2022; Zhixia et al., 2018). Such banks are not concessionary in their financing activities, but adopt procedures to include a minimum amount of environmental (along with social and governance) factors in their financial activities. Indeed, most traditional banks have introduced the practice of "green" banking by actively seeking investment opportunities in environmentally friendly sectors or businesses, through the provision of "green" financial products and services. Although commercial banks may differ from alternative banks in terms of their stated motivations for increasing environmental products, these new developments indicate the beginning of a promising drive toward "green" financial products' inclusion in mainstream banking (Mejia-Escobar et al., 2020; Park & Kim, 2020).

Besides the banking sector, a further set of green investment flows come from institutional investors, including pension funds, sovereign wealth funds, and insurance firms. However, this investor group is still limited to a few lines of green products, such as buying green bonds and sovereign green bonds; this is because they are constrained by a number of hurdles, such as a lack of incentives (market or regulatory) to finance appropriate green projects (Sangiorgi & Schopohl, 2021). Indeed, green investments are generally not included in the relevant benchmarks of the main rating agencies (Escrig-Olmedo et al., 2019). Furthermore, the prevailing regulation requires cautious and conservative investment strategies, which prevent such investors from financing green projects with a poor track record (Mahoney & Mahoney, 2021). International Financial Institutions (IFs), which support the green transformation in various ways represent a promising subsector. For instance, they pioneered new ways of financing green growth by introducing, for example, voluntary commitments to consider the climate risks and carbon footprint of potential investments when making investment decisions (Frisari & Stadelmann, 2015). Furthermore, they pioneered the mobilization and channeling of private and institutional capital for green investments through green bonds (Deschryver & De Mariz, 2020).

Finally, asset managers represent a further main link along the green financial intermediation chain. Asset managers that create investment products—which include both active investment funds and passive funds, such as the Exchange Traded Funds (ETFs)—can be classified according to their investment strategy. These can include screening strategies based on the exclusion of non-environmentally aligned issuers or projects, best-in-class selection, or ethically "incorrect" sectors (Gimeno & Sols, 2020). Other strategies adhere to alignment criteria for issuers/investees, who adopt voluntary self-regulation in terms of ESG factors, for example; or to positive screening criteria that include issuers/investees who proactively seek environmental outcomes from their projects or businesses (Breedt et al., 2019). In terms of these strategies' evolution over time, it can be seen that they moved from management based on prohibiting investment in certain non-ethically compliant sectors, to active management, where the investment choice favors enterprises operating in sectors with a positive social or environmental impact. For these reasons, it is possible to identify three generations of funds: the first is more oriented to the exclusion of investments in unethical industrial sectors; the second has a positive approach to selecting investment targets by adopting social and environmental issues. The third generation aims to reconcile adequate financial returns with support for sustainable development by seeking for positive impact returns.

Finally, the selection of entities in which the fund can invest is based on negative or positive ESG criteria (Amel-Zadeh & Serafeim, 2018). The selection activity may refer to a series of green or ESG ratings; some of these are quantitative, in that they use and weight numerous subcategory metrics based on identified quantitative data, which is either

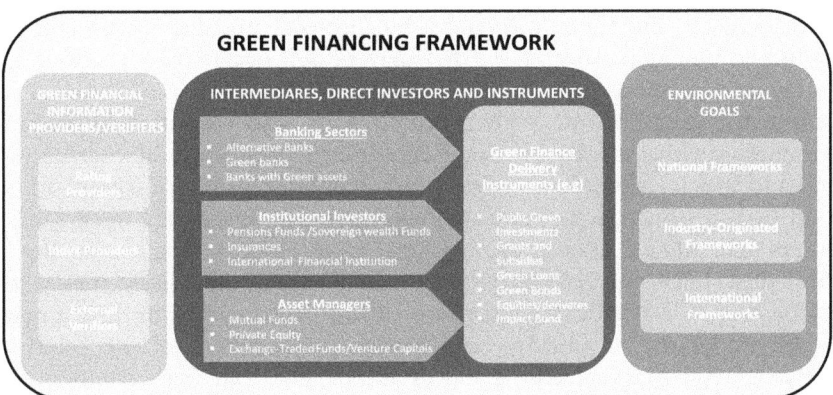

Fig. 3.1 The green financing framework (*Source* Author's elaboration)

offered by corporate issuers or drawn from other sources such as financial media (Verheyden et al., 2016). In many cases, such ESG scoring providers also produce indices to indicate the performance measures for ESG-scored potential investees (or sectors); portfolio managers can use these as a benchmark for investees' ability to generate risk-adjusted returns (Sherwood & Pollard, 2018).

An overview of the green financing framework in the green finance market is represented in Fig. 3.1.

3.3 The Main Green Finance Delivery Instruments: An Overview

Green financial products offer a variety of solutions, and partially represent a mature stage of development. They are essentially segmented into equity, debt, and hybrid instruments.

The most popular and mature debt product is the green bond. Green bonds are fixed-income securities that raise capital for use in projects or activities with specific climate or environmental sustainability purposes. The financial characteristics of such products are similar to traditional or standard bonds in terms of pricing and rating, for example, but they involve proceeds essentially focused on green projects (Gilchrist et al., 2021). A further distinction can be observed among bonds that are *green labeled* or not (Ehlers & Packer, 2017). Green bonds are proceeds earmarked for climate or environmental projects, and have been labeled as "green" by the issuer. The label may be self-declared or certified by an external reviewer. The label requires following precise procedures in issuing the green bond. Such procedures are also identified by issuing standards such as the Green Bond Principles to explain their characteristics, and which recommend a clear process and disclosure for issuers, to be used by investors, banks, underwriters, arrangers, placement agents, and others (Barbalau & Zeni, 2021; Immel et al., 2021). The Green Bond Principles (GBP) are voluntary best practice issuing guidelines; they were established in 2014 by a consortium of the main worldwide investment banks, under the monitoring and project development of the International Capital Market Association (ICMA). The four core components for alignment with the GBP are:

- Use of Proceeds
- Process of Project Evaluation and Selection
- Management of Proceeds
- Reporting.

In order to identify eligible projects for green bonds issuance, the use of proceeds includes a range of activities. In addition to identifying how the proceeds will be used for each bond individually, many green bond issuers develop green bond frameworks in order to illustrate their process for project evaluation and selection. According to ICMA (2021), the categories considered eligible for green bond issuance are listed in Table 3.1.

Finally, different standards have been specified at various levels in order to identify eligible projects. First, the Climate Bond Initiative issued the Climate Bond Standard (CBS), based on GBP, by setting a clear taxonomy of eligible green projects. After external verification of the pre- and post-issuance disclosure of the identified list of projects, it is possible to obtain the CBS certification. Beyond CBS, other international and regional green bond frameworks or guidelines have been issued. Notable among these is the EU Green Bond Standard, which identifies eligible projects as those aligned with at least one of the Environmental Objectives stated in the EU Taxonomy Regulation, while at the same time, not significantly harming any of the others (Marcos & Castrillo, 2022). In addition, the Asian market has two main guidelines: one from the ASEAN

Table 3.1 The use of proceeds in green bonds: the eligible green activities

Eligible green projects category	Examples of projects	Related communication for investors
Renewable energy	Production, transmission, and related technologies	Issuers sould provide information about: • The environmental sustainability objectives of the eligible green projects; • The way by which the projects fit well with eligible categories; • The evaluation and measures adopted for the environmental risks associated to the eligible projects
Energy efficiency	Energy storage, smart grids, and district heating	
Pollution prevention and control	Greenhouse gas control, waste prevention, and waste recycling technologies	
Environmentally sustainable management of natural resources	Reforestation, sustainable agriculture, and green aquaculture	
Biodiversity protection	Marine conservation, biodiversity conservation ecosystems	
Clean transportation	Electric vehicles, rail, and multimodal transportation	
Sustainable water/wastewater management	Sustainable infrastructure for drinking water, wastewater treatment, and flooding mitigation projects	
Climate change adaptation	Technology for climate change monitoring, infrastructures to improve resilience to climate changes	
Circular economy	Reused materials, circular tools and services, design of recyclable materials	
Green buildings	Certification of energy performances, green lighting	

Source Author's elaboration

Capital Market Forum in 2018, which does not include a specific list of green areas (Azhgaliyeva et al., 2020). Furthermore, in June 2020, the People's Bank of China created the *China Green Bond Endorsed Project Catalogue and the Standard*, which provided an official list of eligible green areas (Zhang, 2020).

Green bonds can be classified by issuer type or by debt form. The first case includes sovereign or supranational green bonds, local or municipal green bonds, financial or non-financial corporate green bonds, and multilateral development banks' or commercial banks' green bonds (Banga, 2019; Clapp & Pillay, 2017; Lebelle et al., 2020; Wiśniewski & Zieliński, 2019). The sovereign green bond represents the most rapidly growing green bond segment (Maltais & Nykvist, 2020). According to the Climate Bonds Initiative (CBI) (2022), among sovereign green bonds, in 2021 the United States and France were two of the top four biggest issuing countries of green bonds, with Germany and China growing, respectively, to be the second and fourth largest. The second form of classification distinguishes the use of proceeds structured green bond and environmental/sustainable (and general) corporate purposes linked bond. Regarding the market trend, the green bond market has expanded strongly in the years following the Paris Agreement of 2015. According to the Climate Bond Initiative (2021), the green bonds market reached USD 354.2bn at end of September 2021, surpassing 2020s total.

Finally, green equity instruments refer to equity investments or shares that promote environmental sustainability, such as investments in private equity funds that support renewable energy projects (Taghizadeh-Hesary & Yoshino, 2020). Such green finance delivery instruments are variegated and in a rapidly evolving context (Konisky & Carley, 2021). In particular, a leading role is played by green equity investments by publicly backed financial institutions—both development finance institutions (DFIs) and dedicated multilateral climate funds (Buller, 2022) —which mobilize private finance for green benefits, whether from commercial banks or private equity funds.

Furthermore, green private equity funds, especially those focusing on renewable energy, registered a significant growth since 2015, the year of the Paris Agreement. However, according to the International Monetary Fund (2021), the sector is too limited in size and scope, compared to the total amount of equity investments, to have a major impact among green finance delivery instruments, and it faces challenges related to greenwashing. Indeed, green equity products exist in great variety throughout the market, using a wide range of approaches to green investing. A predominant portion of funds and institutional mandates are managed passively, such as by tracking a reference index very closely (Hoque, 2020). In recent years, many indices have been developed to identify and track the performance of green industries, but there remains a certain degree of opacity in the methodology used to build the index (Sahin et al., 2022). Thus, the true scale of green equity investments depends very much on the definition of "green." Hence, an overview of the main green index providers is reported in Table 3.2.

Finally, beyond the above-described segmentation, as regards the banking sector, according to Akomea-Frimpong et al. (2021), the main green products and services delivered are:

- Green credit/loans
- Green long-term investment accounts
- Carbon finance
- Climate finance
- Green traded stocks and bonds
- Green bancassurance
- Green infrastructural finance

Table 3.2 Main index providers in green finance, approach and some examples

Investment approach

	Positive/best-in-class	Negative screening	Thematic investing
MSCI	MSCI Global Low Carbon Indexes	MSCI Global Fossil Fuels Exclusion Indexes	MSCI Global Environment Indexes MSCI Global Climate Indexes
FTSE	FTSE4Good Index Series	FTSE All world excluding fossil fuels/coals Index Series	FTSE Environmental Opportunities/Technology Index Series
DOW JONES	Dow Jones Sustainability World Index—Diversified Indexes	Dow Jones Sustainability World Index—Diversified Select Indexes	Dow Jones Brookfield Global Infrastructure Net Zero

Source Author's Elaboration

A green loan is a form of financing that enables borrowers to use the proceeds to exclusively fund green innovation and projects. Principles and standards have also been created for the delivery of such products. These are mainly represented by the Green Loan Principles of the International Capital Market Association (ICMA), and by the Equator Principles, which encourage banks to increase the amount of loans to be disbursed in green projects by reducing the interest on these loans, as well as extending the repayment period (Li et al., 2018). Also in the case of green loans, green long-term investment accounts are established to fund eligible green projects (Deka, 2015). Carbon and climate finance are intended as funds to target specific environmental goals, such as the reduction of greenhouse gases, and activities and projects for climate change adaptation and mitigation.

Among green instruments that are not debt-based, further green products from banks are represented by green traded stocks; green bancassurance, which offers carbon–neutral underwriting coverage for green buildings or cars; and, finally, green infrastructural finance, which is a typical green instrument for development projects (Bhattacharyya, 2022).

3.4 The Emerging Green Finance Markets and Instruments

Within the green finance sector, the three main segments of financial products have been illustrated. *Green bonds, green loans,* and *equity investments* (largely performed with indices) represent the main avenues for those capitals that are intended to address environmental benefits. However, further solutions are emerging from various forms of financial innovations; these are intended to update capitalism through innovation, in order to succeed in achieving society's goals (Shiller, 2013). In more detail, three main innovations have expanded the toolbox of green finance instruments: *impact bonds, crowdfunding,* and *sustainability-linked bonds/loans.*

Green Crowdfunding

Classified among the alternative finance innovations, green crowdfunding (GCF) facilitates the financing process for projects, people, and organizations, through an online platform that enables retail investors and individuals (the "crowd") to support green initiatives, green-oriented ventures, and small and medium environmentally friendly enterprises, through investing small amounts of capital (Tenner & Hörisch, 2021). It is worth noting that crowdfunding represents an emerging form of "disintermediated" investments (Cai, 2018), given that a single investor directly collects information and makes decisions regarding the destination of their capital. In this context, the motivations of crowdfunders can be influenced, beyond the classical financial parameters of analysis, by emotional stimulus from the aim of the project itself, and its environmental potential claimed in the crowdfunding campaigns (McLaren & Baldegger, 2021).

By bringing like-minded investors and green entrepreneurs together, crowdfunding can help to scale-up green innovations developed by environmental entrepreneurs (Dudin et al., 2019), and in this way contribute to achieving green growth on a global scale; even though the dynamics of green crowdfunding are much more complex than in conventional crowdfunding (Testa et al., 2019). The growing number of green-focused crowdfunding platforms reflects this trend.

The main categories of crowdfunding platforms present in the green finance sector are lending-based and equity-based. In the first, the crowdfunding platform enables the matching of borrowers and lenders, sometimes by adopting a thematic approach (for instance, energy sector or civic projects). In the second form of platforms, the green entrepreneurs sell small ownership stakes in their firms, generally of a small amount. In this case, crowdfunders invest in green startups that aim to enter the market, or small and medium enterprises seeking capital to launch new businesses or to expand their business model (Vismara, 2019). Within the green finance sector, the segment of equity-based crowdfunding is the most innovative and highly developed.

Beyond the distinction between loan- or equity-based platforms, the potentials of such alternative channels for green projects and activities have been identified both in academia (among others, Appiah-Otoo et al., 2022; Bento et al., 2019; Böckel et al., 2021; Lam & Law, 2016; Martínez-Climent et al., 2019), and by green market-building institutions (see, for instance, Picón Martínez et al., 2021, or World Bank, 2013). However, it is notable that these innovative tools have attracted two main green sectors: energy (especially renewable energy), and green building sectors. A list of the leading green crowdfunding platforms is presented in Table 3.3.

Table 3.3 List of the leading crowdfunding platforms for green investments

Platform	Minimum investment (USD)	Green investment options	Financial notes
WeFunder	100	The platform offers a wide projects' selection for cleantech, sustainability, social impact, and energy companies	Investments average returned 5.9 × the initial investment (data disclosed for the period 2013–2016)
StartEngine	100	The platform offers a wide projects' selection that can be filtered by industry, including environment and clean technology	The platform offers a secondary trading marketplace
Republic	10	The platform offers a wide projects' selection of green projects but not categorized	The platform is co-investor owning a 2% of all securities raised
AngelList	1000	The platform offers multiple options of green investment opportunities (funds and deals)	The platform only accepts accredited investors; investment options are: behind venture capital fund managers (Rolling funds) or in a broad venture capital fund (AngelList access fund)
Fundable	1000	The platform offers a wide projects' selection (both equity and rewards-based) but not categorized	The platform doesn't charge investors any fees. In contrast, companies pay a flat monthly fee to be listed on the platform
Raisegreen	100	The platform exclusively offers investments in green energy projects. So the selection offers not wide alternatives	The platform is focused on impact investing approach
NetCapital	99	Wide selection of green projects (both equity and rewards-based) but not categorized	The platform offers a secondary platform for potential liquidity
SeedInvest	500	Limited selection of green projects (the platform offers multiple categories of projects)	The platform has since raised over USD 300 million for over 235 startups

Platform	Minimum investment (USD)	Green investment options	Financial notes
Microventures	100	Limited selection of green projects (the platform offers multiple categories of projects)	The platform offers a secondary trading platform

Impact Bonds

Among the investment strategies aiming to achieve sustainability outcomes, *impact investing* has shown a rapid growth. This investment strategy aims to simultaneously achieve positive financial benefits, as well as measurable social or environmental returns (Bugg-Levine & Emerson, 2011). Beyond the differences relating to particular geographical regions and sectors, according to the Global Impact Investing Network (GIIN) (2019), impact investing has four main characteristics:

- *Intentionality*: Impact investors strategically invest with the intent to have a positive social or environmental impact;
- *Expectation of Financial Returns*: When impact investors assess their deals they expect a financial return on their capital, or, at minimum, a return that can range from below market to risk-adjusted market rate;
- *Range of Asset Classes*: Impact investments can be made across asset classes;
- *Impact Measurement*: The social and environmental performances of underlying investments need to be measurable.

Impact investing can have various non-financial returns, and, when directed to achieve environmental objectives, it is also called "environmental impact investing" (Birindelli et al., 2020). One of the most innovative and growing impact investment financing instruments is *impact bonds* (IB), which are useful for achieving social as well as environmental objectives. According to Outes Velarde et al. (2022), since its first application in the United Kingdom in 2010, until the beginning of January 2022, the Impact Bond Dataset of the Government Outcomes Lab of Oxford University identified 226 impact bond projects worldwide. However, only a limited percentage of them were deployed for environmental outcomes, defined as *environmental impact bonds* (EIBs). Nevertheless, IBs' potential to bridge the funding gap for environment-oriented projects has been widely recognized (Balboa, 2016; Bose, 2020; Brand et al., 2021; Chen & Bartle, 2022). Impact bonds include both public and private actors, who are mainly:

- *Investors*, who provide working capital for environmental projects and are seeking to be repaid the original investment with a premium;
- *Commissioners*: Entities that receive benefits from the project enabled by the EIB, and repay EIB principal and interest as those benefits accrue;
- *Project delivery entities*, who implement the environmental interventions on the ground.

Other actors involved in an IB are (*a*) independent evaluators, who measure and certificate the achievement with a precise measure of impact; and (*b*) impact-finance specialized intermediaries, who facilitate the design of the project, the contractual issues, or the financial flows among IB actors (Fraser et al., 2018).

The IB financing scheme is based on a pay-per-outcomes rationale (Edmiston & Nicholls, 2018). In other terms, investors will receive their initial investment only if the project achieves an ex-ante determined outcome threshold (which means the "success" of the investment).

A functioning scheme of an EIB is represented in Fig. 3.2.

To conclude, EIBs are a promising financial structure, but frameworks regarding environmental impact and accountability must be carefully improved as the field expands (Balboa, 2016; Dey & Gibbon, 2018), through additional research which could support the development and standardization of such financing tools.

Sustainability-Linked Bonds

Sustainability-linked bonds (SLBs) have been defined as a type of bond instrument whose financial and/or structural characteristics show a forward-looking non-financial performance logic. More simply, SLBs can vary depending on whether the issuer achieves the predefined sustainability or ESG objectives. Thus, the coupon payment is linked to the issuer's sustainability performance (e.g., lower greenhouse gas emissions). Unlike traditional green bonds, a sustainability-linked bond does not consider the use of proceeds; instead, the issuer sets key performance indicators (KPIs) which are aligned with their sustainability strategies. This allows the issuer to set more general, overarching sustainability goals, rather than being tied to financing specific projects. Thus, SLBs are issued under the condition that the issuer will improve their overall sustainability performance, independently from financing investments in green-labeled

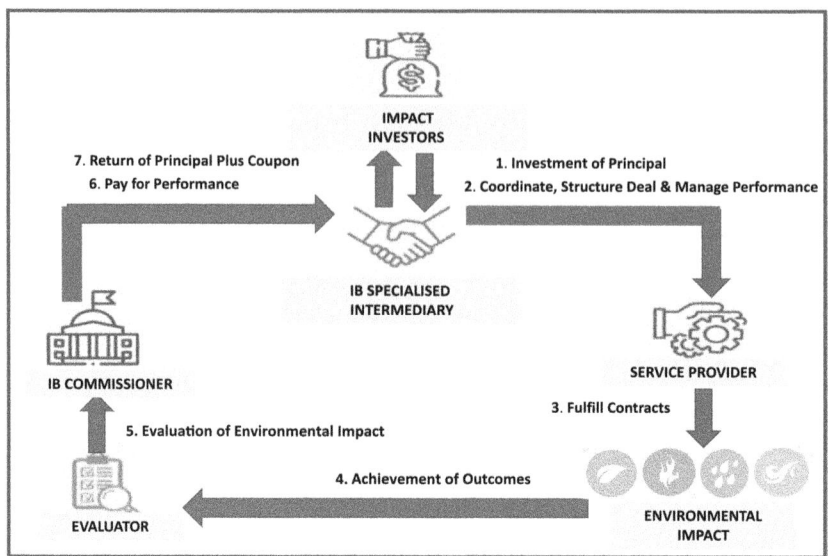

Fig. 3.2 The Environmental Impact Bond model (*Source* Author's elaboration)

projects. However, although such flexibility allows a broader universe of issuers to obtain sustainable financing without the related cost of issuing a green bond, it may also encourage greenwashing practices (Broadway, 2022). Indeed, such characteristics are particularly useful for high-yield issuers, who are typically companies with smaller capital structures, and which might struggle to identify enough green investment projects that can qualify for a green bond issue. To confirm this point, companies issued more than 40% of the total SLB supply in 2021 with a high-yield rating (Bloomberg, 2022). Since the issuance of the first SLB by the Italian utility company Enel in 2019, more than USD 105 billion of these bonds have been issued in 2021, representing 15% of all ESG-themed bond issuance (Bloomberg, 2022).

Beyond greenwashing concerns, SLBs have a further critical problem in their original financial scheme, that could bring moral hazard perils. This issue is related to the SLB financial structure, which includes a coupon step-up feature that increases return payments if the issuer fails to meet the predefined sustainability objectives (Liberadzki et al., 2021). A schematic flow of SLBs is illustrated in Fig. 3.3.

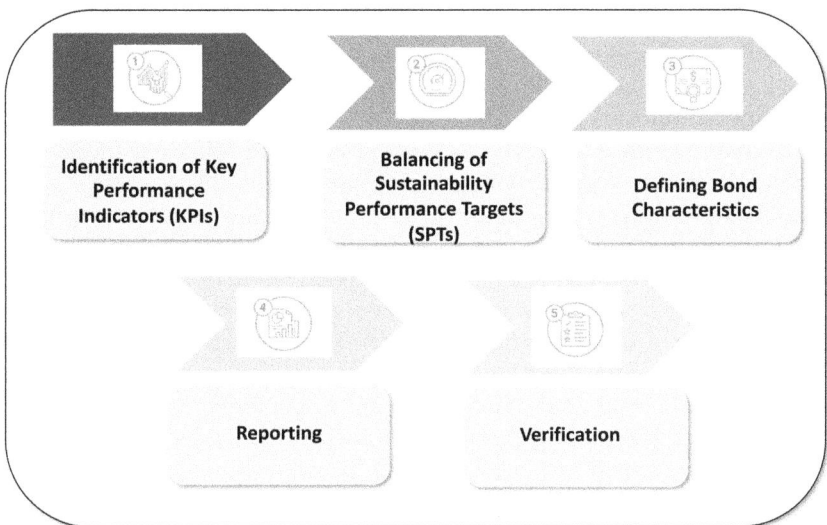

Fig. 3.3 The Sustainability-Linked Bond model (*Source* Author's elaboration)

However, such products are relatively new and, therefore, their potential is doubted by regulators and policymakers, due to the lack of internationally agreed principles that prescribe standardized metrics linked to SLBs (Shrimal, 2021). In terms of regulatory advancements, it is notable that in 2020, the European Central Bank included SLBs among collaterals, while the European Commission added the development of an SLB label to its renewed sustainable finance strategy targets (Tan, 2022). Furthermore, policymakers are also advancing proposals for growing the SLB market even further, by, for instance, linking sovereign debt in developing countries with national sustainability commitments (Giráldez & Fontana, 2022; Vulturius et al., 2022).

3.5 Case Studies of Emerging and Innovative Green Financing Tools

This section provides an overview of case studies to illustrate each of the emerging and innovative green finance tools, as highlighted above, in the fields of crowdfunding, impact bonds, and sustainability-linked bonds. Respectively, these are the case of Raisegreen, one of the main relevant green equity crowdfunding platforms in the United States; the Forest Resilience Impact Bond, concerning the Tahoe National Forest in California; and the Sustainability-Linked Bond issued by Imerys, a French multinational mineral-based firm.

The crowdfunding platform Raisegreen is a green-focused investment marketplace for environmental impact investors who invest in debt notes or equity stakes in green projects, with a minimum investment determined by the project's presenter. Individuals, startups, non-profits, or even local governments can present their offering on Raisegreen's platform for potential investors to start their own green projects. The platform offers investment opportunities in a wide variety of green sectors, such as solar power, affordable housing, sustainable agriculture, or water projects, which have a broad green target of tackling climate change. Unlike other green crowdfunding platforms, Raisegreen focuses on the impact of projects presented on the platform by performing a form of due diligence process which assesses the project's revenue, ambition, environmental impact, and related metrics. Furthermore, the platform assists project promoters by providing legal as well as financial assistance; it therefore represents the ideal finance-raising tool for projects that may have powerful local impacts, especially for disadvantaged communities. Indeed, the platform helps communities by lowering barriers to entry for new project developers, such as high upfront capital costs, high-interest rates on loans, and lack of verifiable impact certification due to the scarcity of community ownership of projects and a lack of usable metrics. Thus, adopting this business model allows a reduced risk profile of investments presented on the platform, and a lower assessment of the capital outlays projects would need. The phase of security structuring, of a maximum of 18 months, enables the investor to see profits in a relatively short time.

As an example of profitable green projects funded through the platform, the *BlocPower* offer presents several points of interest. The proposal was launched in January 2022 and closed its subscription after few weeks, before the closing date of March 22, 2022. The project offered a minimum principal investment of USD 100, in the form of a climate impact note with a fixed interest rate of 5.50% per year, paid annually for the 12-year term of the note issued by BlocPower Energy Services 3, a project company wholly owned by BlocPower LLC. The campaign raised almost USD 1,000,000, the maximum expected capital amount to fund 61 revenue-generating projects in mid-2022 (Raisegreen, 2022). Specifically, the climate impact notes fund a portfolio of already vetted, and largely completed, energy retrofit projects from BlocPower's pipeline. The projects' expected impact spans three outcomes, which primarily benefit American communities where the projects are implemented:

- Improving public health, thanks to reducing the spread of pollutants and infectious diseases, including COVID;
- Addressing climate change by facilitating the low-carbon transition, and also improving energy efficiency, which will reduce greenhouse gas emissions;
- Improving economic outcomes by reducing energy bills for tenants and increasing net operating income for building owners, and by generating employment opportunities.

BlocPower adopted an internationally recognized, auditable system to measure the carbon reduction, tracked by a blockchain-based system. Projects financed by this project are expected to have long-term 15–20 year contracts and relative revenues from 15–20 year leases, while building customers for completed projects to support payback: for instance, an amortizing mortgage at 5.5% interest, plus the original investment returned.

The *Forest Resilience Impact Bond* (FRIB) was launched by Blue Forest Conservation, a mission-driven, non-profit organization specialized in green financial sustainable solutions; the pilot impact bond was offered in 2018. The FRIB is USD 4 million in size and located in the Tahoe National Forest, California.

The project started from the recognition of private capital's role in bridging the growing financing gap, between the estimated USD 60 billion needed to restore forests within the United States, and the Forest Service's annual average budget of USD 500 million allocated annually. In more detail, forest restoration is required to reduce the risks associated with unnaturally dense forests, as a consequence of a lacking forest restoration activities due to progressively severe and costly wildfires that diverted funds to fire suppression, away from prevention through forest restoration.

The FRIB adopted financing for forest restoration by recognizing that, if contracted appropriately, it was an opportunity for private investors to provide upfront investment to cover the initial costs of restoration. One of the innovative aspects of the FRIB was the identification of various stakeholders who draw economic value from forest restoration efforts, and could thus share the cost of the project. The partners of the project are the National Forest Foundation (NFF) as an implementing partner, and the Yuba Water Agency (YWA) and the Forest Service (payors). It was financed with USD 4 million from two concessional investors, the Rockefeller Foundation and the Gordon & Betty Moore Foundation, as well as two commercial investors, CSAA Insurance and Calvert Impact Capital. All investors have a similar seniority ranking, and receive principal and interest on a quarterly basis, at an interest rate of 1.0% per annum for concessional investors and 4.0% per annum for commercial investors.

Regarding the expected impact, forest restoration activities started in 2019 and are expected to be completed by 2022, yielding considerable environmental benefits through reducing the incidence of large damaging forest fires. Specifically, the expected environmental impact was expressed in the following forms:

- High-severity wildfire risk reduction (and of the related costs associated with fire suppression);
- Maintaining water quality and preventing sedimentation due to wildfire;
- Additional water quantity deriving from reduction in forest water use;
- Reduced impact of floods and the resulting costs;
- Carbon sequestration and avoiding air pollution emissions;
- Rural community development.

The peculiarity of the project is that the payments were negotiated on expected outcomes, with beneficiaries providing contracted project cash flows in an arrangement similar to infrastructure financing. In particular, investors found sufficient value for their financial return in the FRIB's ability to accelerate the scale of restoration work that could be achieved due to the availability of upfront funding.

The FRIB project serves as a pilot for scaling-up the funding of projects to the USD 10–25 million range for larger areas of restoration work, by demonstrating its commercial viability and ability to accelerate the pace of ecological work. Therefore, impact bonds represent a multi-stakeholder financing model that can be applied as a replicable model for the market to many environmental problems, in both developing and developed contexts.

Imerys' Sustainability-Linked Bond was introduced on May 14, 2021, for a total amount of €300,000,000. The bond will bear interest at the rate of 1.00% per annum. Moody's and Standards & Poor's, based on Imerys' financial situation, and on the bond's financial characteristics, rated it respectively Baa-3 and BBB-. In case of a rating downgrade, the bond scheme is structured to increase the original rate of interest in accordance with the specific level of the lower rating: up to 3.5% per annum in case of a downgrade to B3 and B- or lower. If Imerys does not achieve the sustainability performance target by the date determined by the issuer, based on greenhouse gas (GHG) emissions intensity as reviewed by an external verifier, Imerys will pay an amount equal to 0.25% of the principal amount for each bond (defined as the Premium Payment Amount). The expected sustainability targets are expressed in terms of a reduction of the GHG emissions intensity equal to or higher than 22.9% by December 31, 2025, compared to the 2018 baseline. Furthermore, a second target observation date is scheduled, with an expected reduction in GHG emissions intensity equal to or higher than 36% by December 31, 2030, compared to the 2018 baseline. The GHG emissions intensity is the sum of direct greenhouse gas emissions of the Imerys group, expressed in tons of carbon dioxide equivalent, per million euros of revenue of the group, as reported in its Sustainability Performance Reporting. Imerys' sustainability-linked bond framework was established in accordance with the five core components of the Sustainability-Linked Bond Principles 2020, published by the International Capital Markets Association:

- Selection of Key Performance Indicators;
- Calibration of Sustainability Performance Targets;
- Specific characteristics of the bonds;
- Reporting;
- Verification.

As the key performance indicator is greenhouse gas emissions reduction, Imerys has therefore selected GHG emissions intensity as a measure. Hence, unlike green bonds, which focus only on the proceeds-based financing of eligible projects, sustainability-linked bonds address general-purpose financing, under the condition that issuers make material and ambitious improvements to their overall sustainability performance.

3.6 Conclusion

Green crowdfunding, impact bonds and sustainability-linked bonds undoubtably have the potential to encourage investments in green and low-carbon activities. These help to reduce the financial gap in aligning actions with the trajectory of the Paris Agreement, and provide capital for economic activities that aid the transition. However, several issues affect the ability of such instruments and marketplaces to realize their potential to accelerate the decarbonization of the real economy at the required speed. First, science-based targets should become best practice for the selection and identification of environment-related performance indicators and targets that are adopted in projects financed with these innovative tools. For instance, such measures should ensure that investors and market observers can easily check and compare the environmental performance of the investment against their science-based climate targets, while simultaneously reducing potential adverse effects for the credibility of the respective markets. Furthermore, actors in these markets need to reach a common understanding of economic activities that materially contribute to environmental targets, as this could help market participants in setting material performance indicators and ambitious targets.

In conclusion, financial innovation in the green finance sector is moving toward the adoption of instruments that facilitate the tracking of environmental performance—in some cases expressed in terms of environmental impact—beyond a simple taxonomy-based identification of green activities. This allows a significant improvement in the accountable delivery of green finance, to more quickly reach the environmental targets fixed by Paris Agreement. However, efforts are also required from researchers interested in the field, to reduce the complexities of procedural or structuring models. Thus, this overview provides insights for the next three chapters, and permits a detailed investigation of issues concerning the identification and accountability of environmental returns in green investments.

References

Akomea-Frimpong, I., Adeabah, D., Ofosu, D., & Tenakwah, E. J. (2021). A review of studies on green finance of banks, research gaps and future directions. *Journal of Sustainable Finance & Investment*, 1–24.

Amel-Zadeh, A., & Serafeim, G. (2018). Why and how investors use ESG information: Evidence from a global survey. *Financial Analysts Journal, 74*(3), 87–103.

Appiah-Otoo, I., Song, N., Acheampong, A. O., & Yao, X. (2022). Crowdfunding and renewable energy development: What does the data say? *International Journal of Energy Research, 46*(2), 1837–1852.

Azhgaliyeva, D., Kapoor, A., & Liu, Y. (2020). Green bonds for financing renewable energy and energy efficiency in South-East Asia: A review of policies. *Journal of Sustainable Finance & Investment, 10*(2), 113–140.

Balboa, C. M. (2016). Accountability of environmental impact bonds: The future of global environmental governance? *Global Environmental Politics, 16*(2), 33–41.

Banga, J. (2019). The green bond market: A potential source of climate finance for developing countries. *Journal of Sustainable Finance & Investment, 9*(1), 17–32.

Barbalau, A., & Zeni, F. (2021). *The optimal design of green securities.*

Ba, S. (2022). Green finance: Challenges and opportunities. In *The new cycle and new finance in China* (pp. 259–263). Springer.

Bhattacharyya, R. (2022). Green finance for energy transition, climate action and sustainable development: Overview of concepts, applications, implementation and challenges. *Green Finance, 4*(1), 1–35.

Bento, N., Gianfrate, G., & Thoni, M. H. (2019). Crowdfunding for sustainability ventures. *Journal of Cleaner Production, 237*, 117751.

Birindelli, G., Trotta, A., Chiappini, H., & Rizzello, A. (2020). Environmental impact investments in Europe: Where are we headed? In *Contemporary issues in sustainable finance* (pp. 151–175). Palgrave Macmillan.

Biswas, N. (2011). Sustainable green banking approach: The need of the hour. *Business Spectrum, 1*(1), 32–38.

Bloomberg. (2022). *Bonds linked to sustainability goals could become ethical debt of choice.* https://www.bloomberg.com/news/articles/2022-01-31/bonds-with-sustainable-targets-are-flavor-of-year-for-esg-market

Böckel, A., Hörisch, J., & Tenner, I. (2021). A systematic literature review of crowdfunding and sustainability: Highlighting what really matters. *Management Review Quarterly, 71*(2), 433–453.

Bose, S. (2020). Adaptation finance: A review of financial instruments to facilitate climate resilience. In *The Palgrave handbook of climate resilient societies* (pp. 1–23). Palgrave Macmillan.

Brand, M. W., Quesnel, K., Saksa, P., Ulibarri, N., Bomblies, A., Mandle, L., Wing, O., Tobin-de la Puente, J., Parker, E. A., Nay, J., Sanders, B. F., Rosowsky, D., Lee, J., Johnson, K., Gudino-Elizondo, N., Ajami, N., Wobbrock, N., Adriaens, P., Grant, S. B., … Gibbons, J. P. (2021). Environmental impact bonds: A common framework and looking ahead. *Environmental Research: Infrastructure and Sustainability, 1*(2).

Breedt, A., Ciliberti, S., Gualdi, S., & Seager, P. (2019). Is ESG an equity factor or just an investment guide? *The Journal of Investing, 28*(2), 32–42.

Broadway, L. (2022). Should trustees invest in sustainability-linked bonds? *Trusts & Trustees, 28*(4), 275–280.

Bugg-Levine, A., & Emerson, J. (2011). *Impact investing: Transforming how we make money while making a difference*. Wiley.

Buller, A. (2022). The limits of privatized climate policy. *Dissent, 69*(1), 36–42.

Bukhari, S. A. A., Hashim, F., & Amran, A. (2022). Pathways towards Green Banking adoption: Moderating role of top management commitment. *International Journal of Ethics and Systems*.

Cai, C. W. (2018). Disruption of financial intermediation by FinTech: A review on crowdfunding and blockchain. *Accounting & Finance, 58*(4), 965–992.

Chen, C., & Bartle, J. R. (2022). Innovative mechanisms of financing infrastructure. In *Innovative infrastructure finance* (pp. 71–132). Palgrave Macmillan.

Chevallier, J., Goutte, S., Ji, Q., & Guesmi, K. (2021). Green finance and the restructuring of the oil-gas-coal business model under carbon asset stranding constraints. *Energy Policy, 149*, 112055.

Clapp, C., & Pillay, K. (2017). Green bonds and climate finance. In *Climate finance: Theory and practice* (pp. 79–105).

Climate Bond Initiative (CBI). (2021). *Sustainable Debt Summary Q3 2021*. https://www.climatebonds.net/files/reports/cbi_susdebtsum_q32021_03b.pdf

Climate Bond Initiative (CBI). (2022). *$500bn Green Issuance 2021: Social and sustainable acceleration: Annual green $1tn in sight: Market expansion forecasts for 2022 and 2025*. https://www.climatebonds.net/2022/01/500bn-green-issuance-2021-social-and-sustainable-acceleration-annual-green-1tn-sight-market

Deka, G. (2015). Green banking practices: A study on environmental strategies of banks with special reference to State bank of India. *Indian Journal of Commerce and Management Studies, 6*(3), 11–19.

Del Gaudio, B. L., Previtali, D., Sampagnaro, G., Verdoliva, V., & Vigne, S. (2022). Syndicated green lending and lead bank performance. *Journal of International Financial Management & Accounting*, 1–16. https://doi.org/10.1111/jifm.12151

Deschryver, P., & De Mariz, F. (2020). What future for the green bond market? How can policymakers, companies, and investors unlock the potential of the green bond market? *Journal of Risk and Financial Management, 13*(3), 61.

Dey, C., & Gibbon, J. (2018). New development: Private finance over public good? Questioning the value of impact bonds. *Public Money & Management, 38*(5), 375–378.

Dudin, M. N., Ivashchenko, N. P., Gurinovich, A. G., Tolmachev, O. M., & Sonina, L. A. (2019). Environmental entrepreneurship: Characteristics of organization and development. *Entrepreneurship and Sustainability Issues, 6*(4), 1861.

Edmiston, D., & Nicholls, A. (2018). Social impact bonds: The role of private capital in outcome-based commissioning. *Journal of Social Policy, 47*(1), 57–76.

Ehlers, T., & Packer, F. (2017, September). Green bond finance and certification. *BIS Quarterly Review*.

Enoch, C. (2021). The Covid-19 pandemic, the green economy, and financial stability. In *Europe beyond the Euro* (pp. 233–270). Palgrave Macmillan.

Erdelen, W. R., & Richardson, J. G. (2021). A world after COVID-19: Business as usual, or building bolder and better? *Global Policy, 12*(1), 157–166.

Escrig-Olmedo, E., Fernández-Izquierdo, M. Á., Ferrero-Ferrero, I., Rivera-Lirio, J. M., & Muñoz-Torres, M. J. (2019). Rating the raters: Evaluating how ESG rating agencies integrate sustainability principles. *Sustainability, 11*(3), 915.

Fraser, A., Tan, S., Kruithof, K., Sim, M., Disley, E., Giacomantonio, C., Lagarde, M., & Mays, N. (2018). *Evaluation of the social impact bond Trailblazers in health and social care final report* (PIRU Publication 2018–23).

Frisari, G., & Stadelmann, M. (2015). De-risking concentrated solar power in emerging markets: The role of policies and international finance institutions. *Energy Policy, 82*, 12–22.

Gangi, F., Varrone, N., & Daniele, L. M. (2021). Socially responsible banking: Towards a new firm–bank relationship. In *The evolution of sustainable investments and finance* (pp. 101–154). Palgrave Macmillan.

Gilchrist, D., Yu, J., & Zhong, R. (2021). The limits of green finance: A survey of literature in the context of green bonds and green loans. *Sustainability, 13*(2), 478.

Gimeno, R., & Sols, F. (2020). Incorporating sustainability factors into asset management. *Financial Stability Review, 39*, 179–199.

Giráldez, J., & Fontana, S. (2022). Sustainability-linked bonds: The next frontier in sovereign financing. *Capital Markets Law Journal, 17*(1), 8–19.

Global Impact Investing Network (GIIN). (2019). *Core Characteristics of Impact Investing*. The GIIN. https://thegiin.org/assets/Core%20Characteristics_webfile.pdf

Győri, Z., Khan, Y., & Szegedi, K. (2021). Business model and principles of a values-based bank—Case study of MagNet Hungarian Community Bank. *Sustainability, 13*(16), 9239.

Gupta, J. (2015). Role of green banking in environment sustainability—A study of selected commercial banks in Himachal Pradesh. *International Journal of Multidisciplinary Research and Development, 2*(8), 349–353.

Höhnke, N., & Homölle, S. (2021). Impact investments, evil investments, and something in between: Comparing social banks' investment criteria and strategies with depositors' investment preferences. *Business Ethics, the Environment & Responsibility, 30*(3), 287–310.

Hoque, F. (2020). *The Rise of ESG in Passive Investments.* US SIF. https://www.ussif.org/files/Publications/Rise_of_ESG_%20passiveinvestments_2020.pdf

Huang, H., Chau, K. Y., Iqbal, W., & Fatima, A. (2022). Assessing the role of financing in sustainable business environment. *Environmental Science and Pollution Research, 29*(5), 7889–7906.

ICMA. (2021). *Green Bond Principles Voluntary Process Guidelines for Issuing Green Bonds.* https://www.icmagroup.org/assets/documents/Sustainable-finance/2021-updates/Green-Bond-Principles-June-2021-140621.pdf

Immel, M., Hachenberg, B., Kiesel, F., & Schiereck, D. (2021). Green bonds: Shades of green and brown. *Journal of Asset Management, 22*(2), 96–109.

International Monetary Fund. (2021). *Investment funds: Fostering the transition to a green economy.* https://www.elibrary.imf.org/view/books/082/465808-9781513595603-en/ch003.xml

Kocornik-Mina, A., Bastida-Vialcanet, R., & Eguiguren Huerta, M. (2021). Social impact of value-based banking: Best practises and a continuity framework. *Sustainability, 13*(14), 7681.

Konisky, D. M., & Carley, S. (2021). What we can learn from the Green New Deal about the importance of equity in national climate policy. *Journal of Policy Analysis and Management, 40*(3), 996–1002.

Lam, P. T., & Law, A. O. (2016). Crowdfunding for renewable and sustainable energy projects: An exploratory case study approach. *Renewable and Sustainable Energy Reviews, 60,* 11–20.

Lebelle, M., Lajili Jarjir, S., & Sassi, S. (2020). Corporate green bond issuances: An international evidence. *Journal of Risk and Financial Management, 13*(2), 25.

Li, Z., Liao, G., Wang, Z., & Huang, Z. (2018). Green loan and subsidy for promoting clean production innovation. *Journal of Cleaner Production, 187,* 421–431.

Liberadzki, M., Jaworski, P., & Liberadzki, K. (2021). Spread analysis of the sustainability-linked bonds tied to an issuer's greenhouse gases emissions reduction target. *Energies, 14*(23), 7918.

Martínez-Climent, C., Costa-Climent, R., & Oghazi, P. (2019). Sustainable financing through crowdfunding. *Sustainability, 11*(3), 934.

Mahoney, P. G., & Mahoney, J. D. (2021). The new separation of ownership and control: Institutional investors and ESG. *Columbia Business Law Review*, 840.

Maltais, A., & Nykvist, B. (2020). Understanding the role of green bonds in advancing sustainability. *Journal of Sustainable Finance & Investment*, 1–20.

Marcos, S., & Castrillo, M. J. (2022). Sustainable finance in Europe: The EU taxonomy and green bond standard. In *Handbook of research on global aspects of sustainable finance in times of crises* (pp. 114–130).

McLaren, E. M., & Baldegger, R. (2021). Crowd funders' motivations to support impact-oriented projects. *Journal of the International Council for Small Business, 2*(4), 334–339.

Mejia-Escobar, J. C., González-Ruiz, J. D., & Duque-Grisales, E. (2020). Sustainable financial products in the Latin America banking industry: Current status and insights. *Sustainability, 12*(14), 5648.

Migliorelli, M., & Dessertine, P. (2019). The rise of green finance in Europe. In *Opportunities and challenges for issuers, investors and marketplaces*. Palgrave Macmillan.

Montes, L. (1998). Financing sustainable development in Mexico through alternative banks or "green banks." *Journal of Structured Finance, 4*(1), 67.

Neisen, M., Bruhn, B., & Lienland, D. (2021). ESG rating as input for a sustainability capital buffer. *Journal of Risk Management in Financial Institutions, 15*(1), 72–84.

Outes Velarde, J., Hameed, T., Airoldi, M., Carter, E., Gibson, M., & Macdonald, J. R. (2022). *INDIGO impact bond insights—Second edition, Government Outcomes Lab*. University of Oxford, Blavatnik School of Government. https://doi.org/10.35489/BSG-GOLAB-RI_2022/001

Park, H., & Kim, J. D. (2020). Transition towards green banking: Role of financial regulators and financial institutions. *Asian Journal of Sustainability and Social Responsibility, 5*(1), 1–25.

Picón Martínez, A., Cafferkey, P., & Gianoncelli, A. (2021). *Accelerating the SDGs—The role of crowdfunding in investing for impact*. EVPA. https://evpa.eu.com/uploads/publications/Accelerating_the_SDGs-The_Role_of_Crowdfunding_in_investing_for_impact-_2021.pdf

Raberto, M., Ozel, B., Ponta, L., Teglio, A., & Cincotti, S. (2019). From financial instability to green finance: The role of banking and credit market regulation in the Eurace model. *Journal of Evolutionary Economics, 29*(1), 429–465.

Raisegreen. BlocPower Energy Services 3, Raisegreen, 2022, Climate Impact Note—2nd Offer (2022).

Sahin, Ö., Bax, K., Paterlini, S., & Czado, C. (2022, January 28). The pitfalls of (non-definitive) Environmental, Social, and Governance scoring methodology. *Social, and Governance scoring methodology.*

Sangiorgi, I., & Schopohl, L. (2021). Why do institutional investors buy green bonds: Evidence from a survey of European asset managers. *International Review of Financial Analysis, 75,* 101738.

Sharma, M., & Choubey, A. (2022). Green banking initiatives: A qualitative study on Indian banking sector. *Environment, Development and Sustainability, 24*(1), 293–319.

Sherwood, M. W., & Pollard, J. L. (2018). The risk-adjusted return potential of integrating ESG strategies into emerging market equities. *Journal of Sustainable Finance & Investment, 8*(1), 26–44.

Shideler, J. C., & Hetzel, J. (2021). Financing the Transition. In *Introduction to climate change management* (pp. 139–168). Springer.

Shiller, R. J. (2013). Capitalism and financial innovation. *Financial Analysts Journal, 69*(1), 21–25.

Shrimal, G. (2021). *Transition bond frameworks: Goals, issues, and guiding principles* (Working Paper). Stanford University, Sustainable Finance Initiative, Precourt Institute for Energy.

Singh, S., & Jayaram, R. (2021). Sustainable banking: A systematic literature review. *International Journal of Sustainable Society, 13*(2), 116–128.

Sinha, A., Mishra, S., Sharif, A., & Yarovaya, L. (2021). Does green financing help to improve environmental & social responsibility? Designing SDG framework through advanced quantile modelling. *Journal of Environmental Management, 292,* 112751.

Stern, N., & Stiglitz, J. (2022). The economics of immense risk, urgent action and radical change: Towards new approaches to the economics of climate change. *Journal of Economic Methodology,* 1–36.

Taghizadeh-Hesary, F., & Yoshino, N. (2020). Sustainable solutions for green financing and investment in renewable energy projects. *Energies, 13*(4), 788.

Tan, C. (2022). Private investments, public goods: Regulating markets for sustainable development. *European Business Organization Law Review,* 1–31.

Tan, L. H., Chew, B. C., & Hamid, S. R. (2017). A holistic perspective on sustainable banking operating system drivers: A case study of Maybank group. *Qualitative Research in Financial Markets, 9*(3), 240–262.

Tenner, I., & Hörisch, J. (2021). Crowdfunding sustainable entrepreneurship: What are the characteristics of crowdfunding investors? *Journal of Cleaner Production, 290,* 125667.

Testa, S., Nielsen, K. R., Bogers, M., & Cincotti, S. (2019). The role of crowdfunding in moving towards a sustainable society. *Technological Forecasting and Social Change, 141,* 66–73.

Trapp, C. T., & Kanbach, D. K. (2021). Green entrepreneurship and business models: Deriving green technology business model archetypes. *Journal of Cleaner Production, 297*, 126694.

Verheyden, T., Eccles, R. G., & Feiner, A. (2016). ESG for all? The impact of ESG screening on return, risk, and diversification. *Journal of Applied Corporate Finance, 28*(2), 47–55.

Vismara, S. (2019). Sustainability in equity crowdfunding. *Technological Forecasting and Social Change, 141*, 98–106.

Vulturius, G., Maltais, A., & Forsbacka, K. (2022). Sustainability-linked bonds—Their potential to promote issuers' transition to net-zero emissions and future research directions. *Journal of Sustainable Finance & Investment*, 1–12.

Weber, O. (2014). Social banking: Concept, definitions and practice. *Global Social Policy, 14*(2), 265–267.

Wiśniewski, M., & Zieliński, J. (2019). Green bonds as an innovative sovereign financial instrument. *Ekonomia i Prawo. Economics and Law, 18*(1), 83–96.

World Bank. (2013). *Crowdfunding's potential for the developing world*. World Bank. https://openknowledge.worldbank.org/handle/10986/17626. License: CC BY 3.0 IGO.

Yip, A. W., & Bocken, N. M. (2018). Sustainable business model archetypes for the banking industry. *Journal of Cleaner Production, 174*, 150–169.

Zhang, H. (2020). Regulating green bond in China: Definition divergence and implications for policy making. *Journal of Sustainable Finance & Investment, 10*(2), 141–156.

Zhixia, C., Hossen, M. M., Muzafary, S. S., & Begum, M. (2018). Green banking for environmental sustainability-present status and future agenda: Experience from Bangladesh. *Asian Economic and Financial Review, 8*(5), 571–585.

CHAPTER 4

Green Finance and SDGs: Emerging Trends in the Design of Green Investment Portfolios

4.1 Introduction

Green investing has evolved considerably over recent decades. In the business sector, financial actors have also become an increasingly essential driver of sustainable development and human prosperity (Nykvist & Maltais, 2022). The progressive decline in maximizing the shareholder's interest, in favor of managing the environmental, social, and governance (ESG) issues related to the sustainability of the overall company strategy, operations, and products, has become a strategic priority embedded throughout organizations (Chevrollier et al., 2020). The positive relationship between a company's ESG and financial performance has been deeply explored in recent years (for an overview, see Chouaibi et al., 2021; Friede et al., 2015; Zhao et al., 2018), with positive correlations found between ESG performance, operational efficiencies, and financial performance over the long term (Whelan et al., 2021). Similarly, the financial sector started to benefit by using the market information about companies' ESG performance (Brooks & Oikonomou, 2018).

However, more recently, a large portion of the business sector has started to shift the focus on ESG issues in their management operations and strategy, toward exploring the positive impact return of their products and services on society and the environment (Bocken et al., 2019), which directly assists in achieving sustainability goals. In this regard, a catalyzing influence was the launch of the United Nations Sustainable

© The Author(s), under exclusive license to Springer Nature Switzerland AG 2022
A. Rizzello, *Green Investing*, Palgrave Studies in Impact Finance, https://doi.org/10.1007/978-3-031-08031-9_4

Development Goals (SDGs) in 2015, and recognizing the role of private actors in their achievement (Scheyvens et al., 2016). This call for the private sector was particularly imperative for private capitals, given that the funding from public sources was limited, with an investment shortfall of USD 2.5 trillion in developing countries alone (Doumbia & Lauridsen, 2019). Furthermore, the investment sector is increasingly seeking investment returns by investing in companies that demonstrate this attitude to positively contributing to society, or by seeking investment returns from SDG-aligned themes, such as education or green technology (Johnson et al., 2019). Within this arena, investments closely aligned to *SDG green themes* are registering a rapid growth (Zhan & Santos-Paulino, 2021). Thus, a growing number of sustainability portfolios target environment-related UN SDGs, such as clean water and sanitation (SDG 8), climate action (SDG13), and affordable and clean energy (SDG 7).

Based on these considerations, this chapter surveys the evolution of green investment strategies by focusing on those based on the emerging trend of SDGs. First, the chapter provides an overview of SDGs, their green targets, and the role of private finance in their achievement (Sect. 4.2). Then, it assesses how green investing can be classified, and its different approaches (Sect. 4.3). Among them, the focus on ESG investing forms is also assessed (Sect. 4.4), by outlining the role of ESG ratings and indices in the investment value chain (Sect. 4.5). Section 4.6 discusses the topic of SDG investing in financial services and recognizes this as an emerging driver for successfully building a green investment. Finally, Sect. 4.7 focuses on the evolution of green investment strategies based on the discussed results.

4.2 Sustainable Development Goals: Green Targets and Financial Needs

In 2015, all the Member States of the United Nations adopted the 2030 Agenda for Sustainable Development (United Nations, 2016). The initiative is a follow-up of two precedents: a Vision of The World We Want, and the Millennial Development Goals (MDGs), inaugurated in 2000 (Linnér & Selin, 2013). The 2030 Agenda, unlike MDGs, which mainly focused on poverty and hunger in less developed countries, includes 17 Sustainable Development Goals that are designed to involve a broader set of stakeholders and are applicable to all the world's countries and regions, not only developed economies (Arora & Mishra, 2019). The

17 Goals consist of 169 sub-targets and 232 indicators. Their focus can be grouped under three main sustainable development target areas: Economic Growth, Social Inclusion, and Environmental Protection. A graphical representation of this segmentation is provided (Fig. 4.1).

Goals 6, 7, 13, 14, and 15 are directly related to environmental sustainability. However, it is important to note that the environmental outcomes were extensively incorporated within all the SDGs: for example, by placing the most important ones under "non-environmental" goals, such as SDG 8 or SDG 11, considering that economic growth can be made environmentally sustainable using resource efficiency and key green technological solutions (Elder & Olsen, 2019). Thus, the SDGs relevant to environmental protection and sustainability are described in Table 4.1, by specifying the relevant targets involved.

The implementation of the 17 SDGs requires the mobilization of trillions of dollars in resources; this calls for action not only by all governments and public institutions worldwide, but also by business and investors. Indeed, the annual financing gap has been estimated at USD 2.5 trillion annually (United Nations, 2018), and SDGs have started to interest a growing group of investors in supporting businesses that address them. Indeed, the fulfillment of SDGs implies a global growth, in terms of creating new industries or assisting business sectors' transition toward

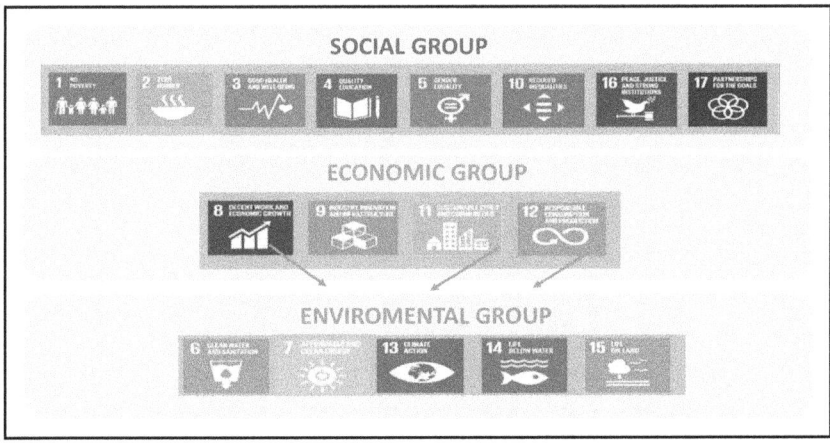

Fig. 4.1 A segmentation of SDGs by broader areas (*Source* Author's elaboration)

Table 4.1 Relevant SDG Targets related to the Environment

SDG	SDG Targets
SDG 1—No poverty	1.5 Resilience to climate and environmental shocks and disasters
SDG2—Zero Hunger	2.5 Promoting/protecting genetic diversity
SDG3—Good Health and Well Being	3.3 Deaths and illnesses from pollution 3.9 Water-borne diseases
SDG4—Quality education	4.7 Education for sustainable development
SDG5—Gender quality	5.a Women's equal rights to economic resources, property, and natural resources
SDG6—Clean water and sanitation	All
SDG7—Affordable and clean energy	All
SDG8—Decent work and economic growth	8.4 Resource efficiency, decoupling economic growth from environmental degradation 8.9 Sustainable tourism
SDG9—Industry, innovation, and infrastructure	9.1 Sustainable and resilient infrastructure 9.2 Sustainable industrialization 9.4 Sustainability upgrading and resource efficiency 9.a Financial, technical, & technological support for sustainable & resilient infrastructure
SDG10—Reduced inequalities	–
SDG11—Sustainable Cities and Communities	11.2 Sustainable transport 11.3 Inclusive and sustainable urbanization 11.4 Protect & safeguard cultural & natural heritage 11.6 Environmental impact, air quality, waste management 11.7 Green and public spaces 11.b Integrated policies on inclusion, resource efficiency, climate mitigation & adaptation, resilience, disaster risk management 11.c Support for sustainable & resilient buildings

SDG	SDG Targets
SDG12—Responsible consumption and production	12.2 Sustainable management and use of natural resources 12.3 Halve global per capita food waste 12.4 Responsible management of chemicals and waste 12.5 Substantially reduce waste generation 12.6 Encourage companies to adopt sustainable practices and sustainability reporting 12.8 Promote universal understanding of sustainable lifestyles
SDG13—Climate action	All
SDG14—Life below water	All
SDG15—Life on land	All
SDG16—Peace, justice, and strong institutions	—
SDG17—Partnerships for the goals	Indirectly related when applied to green SDG targets

Source Author's elaboration

achieving the goals. This is particularly true for environmentally related SDGs, which have opened broad market opportunities in the fields of renewable energy or climate action; for example, by offering investment opportunities with competitive returns (Trabacchi & Buchner, 2019). Corporate sustainability leaders are also identifying business value from SDG- and environment-aligned strategies. Therefore, the private sector has started to determine which SDG impacts are important for their business and investments, and to measure them by optimizing business strategies that align financial value with environmental value (ElAlfy et al., 2020; Johnsson et al., 2020).

4.3 A Framework of Green Investment Approaches

Like any other form of investment, green investment can be made by investing capital in projects, programs, and companies, in the form of equity, debt, or project finance. The use of such forms involves, respectively, buying part of the shares of a green company; lending money, and expecting some interest, for the specific green purpose of a company or private/public organization (generated, for example, by using proceeds or borrowed capitals for a specific green activity). Finally, in project finance, money is directed toward the delivery of a specific green project; international finance institutions such as the European Investment Bank, or other international finance corporations generally adopt this.

All such forms involve different financial risks and different forms of returns, both financial and environmental. Therefore, green investors have a range of different product segments, selection strategies, and appetites for financial and environmental returns (Fig. 4.2).

As shown in Fig. 4.2, the *traditional, responsible, and sustainable* class of investors seek competitive risk-adjusted returns while accepting alignment with green activities/companies, as well as not harming green activities/companies. By contrast, the *emerging and innovative* class of investors present a higher risk tolerance, and may accept concessionary investments while also seeking a positive return on their investments.

Regarding the green targets of investments, green capitals in two main forms can finance green activities, projects, or companies: targeted or untargeted. In the first type, green capitals are directed to specific or known green activities, companies, or projects—by, for example, defining the use of proceeds when issuing a green bond, or in the screening process

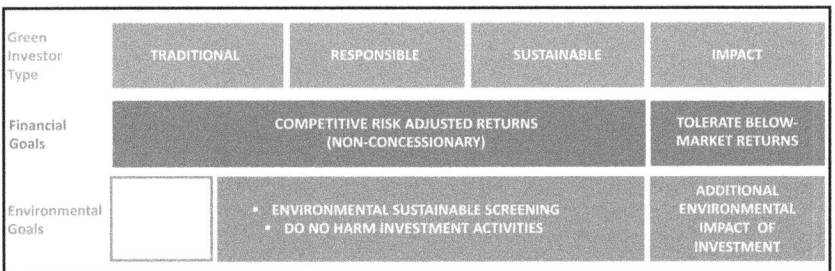

Fig. 4.2 The spectrum of green investors (*Source* Author's elaboration)

for a green loan. In this field, investing capital in financing green activities is significantly assisted by the identification of which sectors, technologies, activities, or projects are considered as "green." In the second class, green investments are not directed at specific green technologies or products, but pursue green targets; therefore, the correlation between the funding and its green impacts is hard to measure. Investors adopting this approach reduce environmental risks across investors' portfolios, as selection or alignment procedures can identify some activities or companies as more environmentally friendly than others. This approach is familiar in listed and private equity investments, as well as in corporate finance. Within this context, activities or companies are identified as "green" if they take a holistic approach to sustainable investment by applying one of the following strategies:

- *Screening*: A strategy for the exclusion of investment in specific sectors, companies, or projects that show, for instance, poor ESG performance (also defined as exclusionary screening). In other cases, the exclusion may involve de-selecting from a portfolio of those companies or projects that do not comply with international norms and standards (norms-based screening). A form of positive screening scheme may involve, for instance, identifying companies that show high ESG performance relative to others (best-in-class screening);
- *Sustainability themed investing*: This investment approach enables the selection of green-theme-focused entities, or investing in assets specifically related to sustainability, such as green real estate, sustainable agriculture technologies, or environmental technology funds.

- *ESG integration*: This approach systematically and explicitly includes ESG factors in the fundamental analysis. Using this strategy, funds are directed to companies that might not be pure players in sustainable sectors, but rather show adequate strategies regarding ESG risk management.
- *Corporate engagement*: Investing in the shares of *grey companies* that use shareholder power to influence corporate "green" behavior or ESG performance;
- *Environmental impact investing*: targeted investments that aim to achieve a positive and measurable environmental impact, alongside financial returns.

Targeted green investment essentially takes place in debt form, consisting of green bonds, green loans, and green infrastructure finance, which were explained in detail in Chapter 3. The focus of this chapter is to explore the main forms of investment strategies regarding untargeted green investments, and to highlight their evolution toward integration with the SDG framework, along with the implications for the green financial sector. Furthermore, this specific segment represents the most debated and controversial field of green finance; therefore, the findings of this chapter will serve as a basis for the analysis in the following two chapters.

4.4 ESG Investing: The Green Investing Giant with Feet of Clay

ESG investing is an approach that systematically incorporates environmental, social, and governance factors within the investment analysis, as well as risk decisions, in order to generate sustainable, long-term financial returns (Halbritter & Dorfleitner, 2015). In other words, this strategy uses forward-looking, financially material information when calculating expectations of returns and risks, for greater long-term benefits (Kotsantonis et al., 2016). ESG differs from traditional investing in that it takes into account the risk assessment factors of long-term environmental, social, and governance challenges and developments, rather than assessing short-term financial performance and the commercial risks of that performance (Krosinsky & Robins, 2012).

The growing importance of ESG investing reflects not only the increased attention to environmental considerations from consumers, investors, and society in general, but also its capacity to generate market financial returns while also improving the risk management of a portfolio (Giese et al., 2019). More simply, the maximization of returns and profits in the financial sector has been progressively accompanied by a need—arising from both risk and opportunity evaluations—to incorporate societal values and broader external factors through a more formalized alignment.

ESG investing has grown rapidly over the past decade, and, according to the Global Sustainable Investment Alliance's (GSIA) biennial review (2021), ESG integration represents the most commonly reported sustainable investment strategy across all regions, registering a total of USD 24.6 trillion AUM. Furthermore, numerous central banks have expressed support for this investment approach by integrating ESG assessment and investing practices into their regulatory or supervisory framework actions (Elsenhuber & Skenderasi, 2020). Regarding the relationship between ESG and financial performance on the investor side, according to the academic review by Whelan et al. (2021), 65% of academic research showed positive or neutral performance compared to conventional investments, with only 13% of academic studies indicating negative findings. The same positive correlation has been also confirmed recently (Broadstock et al., 2021; Chiappini et al., 2021; Pisani & Russo, 2021), in an age of generalized market losses due to the pandemic crisis generated from the COVID-19 spread.

According to Boffo and Patalano (2020), the ESG financing ecosystem includes different actors: first, the issuers of ESG data, who receive ESG ratings from rating providers. A second level of ESG data elaboration is represented by ESG indices. Both kinds of ESG information, ratings and indices, are managed along the financial intermediation chain by asset managers and by institutional investors, who construct and market ESG financing tools and manage assets in order to give the end investors the ultimate rewards and risks of ESG-based investments. In order to offer disclosure guidance and good practices to other actors, the landscape also includes framework developers and providers, such as disclosure standard setters, and exchange and self-regulating bodies.

The ESG integration includes the following main steps:

- Assessment of the ESG policies and processes of companies, and evaluation of those best managing these issues;
- Identifying which issues are material to companies' financial prospects;
- The final process of overweighting or underweighting the companies;
- Assessing related ESG risks;
- Selection of ESG investment targets.

During the described process, the use of ESG ratings, and the derived ESG indices, represents one of the key ways in which investors and other market participants make use of ESG information. However, a wide range of rating practices exist, and the methodologies adopted by rating providers are intrinsically different (Barman, 2018; Berg et al., 2019). Moreover, it is important to note that not all ESG scorings relate to companies which score highly on ESG issues that are financially material to their business (Bender et al., 2018). Thus, current analytical efforts are being made to extract ESG scoring information of materially relevant financial value for those investors seeking absolute and risk-adjusted returns (Cort & Esty, 2020).

However, a major shortcoming of studies on ESG effects is that disclosures are voluntary, and there are no uniform standards to ensure comparability of those items across firms (Rajesh & Rajendran, 2020). Complexity results from the different frameworks, measures, key indicators and metrics, data use, and qualitative judgments adopted in the scoring models. The lack of agreed standards for ESG ratings is one of the main challenges for the sector (Dimson et al., 2020), and raises relevant questions regarding greenwashing risks (Gyönyörová et al., 2021). Thus, the entire ESG data and analytics industry provides information that is not sufficient for assessing these companies' impact in reality (Porter et al., 2019), because selection focuses on the level of responsibility or sustainability involved in the production of good/services, rather than *what* was produced, and the contribution of such goods and services to society.

4.5 The Role of ESG Ratings and Indices in the Investment Value Chain

ESG ratings and indices are a crucial component, allowing investors to better evaluate their portfolio companies' ESG performance. Consequently, as the market for ESG-integrated investment has grown in recent years, these market tools have expanded in quantity, complexity, and variety (Billio et al., 2021; Deng & Cheng, 2019; Reiser & Tucker, 2019). Moreover, it is possible to identify two main ESG scoring approaches: a focus on ESG disclosure, i.e., assessing how the company reports its information regarding sustainability; and a focus on ESG risk management, which assesses the main risks the company faces in terms of ESG (Dorfleitner et al., 2020; Walter, 2020). Thus, portfolio analysis and risk-adjusted returns can be influenced by the choice of ESG providers, timeframes, and location (Gibson Brandon et al., 2021; Vojtko & Padysak, 2019). Hence, different rating methodologies provide very different outcomes, but also an over-simplification, as relying on a single score to assess sustainability, could produce negative financial returns (Avramov et al., 2021; Gibson Brandon et al., 2021).

While ESG ratings refer to a scoring framework to evaluate a listed or privately held company's performance on ESG factors, an ESG index is a list of ESG companies' performances. It ranks them according to the aggregate of individual companies' scores for each pillar: E, S, and G. ESG rating activities are closely related with ESG index construction and, for this reason, many ESG rating vendors are also ESG index providers.

The major current ESG rating and/or index firms are listed in Table 4.2.

In the context of green finance, the use of ESG indices has various applications (Chan et al., 2020; Folqué et al., 2021):

Table 4.2 Major ESG rating and index firms per country

Country	Provider name
Germany	Oekom
The Netherlands	Sustainalitycs
Switzerland	Inrate; RepRisk
UK	FTSE; Trucost
USA	MSCI; S&P Dow Jones; Thomson Reuters

Source Author's elaboration

- Providing performance benchmarks;
- Enabling passive investment strategies based on tracking a particular index;
- Assisting active managers by providing an investment universe to trade in.

Furthermore, an index of companies may be built by examining an ESG issue or set of issues, in order to generate an issue-weighted index (Bender et al., 2017). Alternatively, companies may be included according to eligibility criteria, based on the research that has produced the ESG ratings (Jacob & Wilkens, 2021).

To conclude, the ESG sector is complex and multifaceted. Actors and portfolio strategies rely on different ESG data outcomes, derived overall from companies' sustainability reporting. Thus, the harmonization of ESG rating frameworks and standards will be crucial for the advancement of this sector, in terms of simplifying the sustainability reporting and reducing ambiguity in the related ESG analysis. In this sense, the EU market represents a promising field of improvement areas for the ESG market, due to its recent regulation innovations in corporate sustainability disclosure.[1] Furthermore, the taxonomy of sustainable activities, which helps to achieve the broader targets for climate change, will encourage the expansion of a green-themed investment framework based on the "E" dimension, through the sustainability reports of European companies.

4.6 The Rise of SDG-Aligned Investment Strategies and Their Relevance for Green Finance

After the launch of the United Nations SDGs in 2015, the achievement of the goals has required the growing involvement of the private sector, both in business and finance. Indeed, the urgent call to mobilize capital for SDG achievement has resulted in an increasing number of companies that explicitly and proactively address one or more SDGs through

---

[1] The Corporate Sustainability Reporting Directive (CSRD) is the new EU legislation that, by defining for the first time a common reporting framework, requires all large companies to publish regular reports on their environmental and social impact activities. Compliance is due soon: companies must submit their report aligned with the CSRD on January 1, 2024, for the 2023 financial year. For further information: https://eur-lex.europa.eu/legal-content/EN/TXT/?uri=CELEX:52021PC0189.

their products or services, and report on their efforts. Also, investors have begun to align their investments and impact goals with SDGs, or require investee companies to measure their SDG impact performance (Muhmad & Muhamad, 2021). This emerging trend has added a layer of analysis to existing prominent investment approaches, such as exclusionary screening and ESG integration, and is also a new investment segment that is "impact-aligned" to the SDGs (De Franco et al., 2021).

However, an investment approach, which focuses on achieving SDG outcomes, requires an analysis beyond that of the financial materiality of ESG issues, which refer to an individual investee level, in order to sufficiently capture the outcomes for society and the environment at a systemic level (Madaleno et al., 2022). To overcome the ESG's "partial" approach to capturing SDG impact, it is crucial to understand how investors can achieve outcomes that best align with SDGs, among investments that provide equal financial opportunities. Furthermore, investors need to develop investment frameworks to identify alternative sources of return that can provide SDG-aligned outcomes over a longer time horizon (Chew & Childe, 2019; Lee, 2020).

Assessing which investments are best aligned to SDG outcomes requires a shift in investors' perception of their role in shaping outcomes in the real world, both positive and negative, intended or not. Indeed, investors can have an impact, in terms of an additional change in outcomes, in different contexts. An investor can cause outcomes *directly*, in their own business activities, or they can *contribute* to an outcome through an investment activity that facilitates an outcome from an investee. Finally, investors can directly deliver an outcome that is *directly linked* to the product or service of the investee company (Kölbel et al., 2020). Thus, it is important to distinguish between an assessment of outcomes generated from a specific investee company or project, and the assessment of SDG impact-aligned assets. In the first case, the existing and widely agreed framework is useful for investors. For instance, the Impact Management Project provided a framework that helps to define and evaluate impact at the company level, by posing a number of questions ("what?", "who?", "how much?"), the "contribution," and the "risk" associated with the results of its activities (Impact Management Project, 2022). In particular, the contribution to the SDGs measures the *additionality* attributable to the company's activities, while the risk refers to the risk of not meeting impact targets, or ESG risks regarding portfolio management company's alignment with a positive SDG contribution (Liang et al., 2022).

On the other hand, identifying the SDG performance of companies and portfolios best aligned to SDG outcomes is not a simple task. Beyond the case of companies and portfolios linked to a specific SDG sector or target (for instance, investees focused on water sanitation or renewable energy), in the absence of reliable data derived from sustainability reports or ratings, investors can assess the company or portfolio by engaging with companies to encourage high-quality reporting. For instance, their reports could take into consideration specific SDG KPIs and related metrics associated with widely recognized frameworks such as IRIS+[2] and the Global Reporting Initiative (GRI).[3] In the context of green finance, this discourse is relevant to green SDGs, as detailed in Sect. 4.2. Thus, it is possible to differentiate SDG-based *green thematic investing*, *impact-aligned investing*, and *impact investing*. In more detail, green thematic funds or impact-aligned funds can expose their clients to particular green themes, such as water, renewable energy, or climate change, which are aligned to specific SDG targets, and can require companies to report explicitly on their contribution to specific SDG green goals. In this light, the emerging SDG-aligned listed equity solutions correspond with this trend (Lassala et al., 2021). Accordingly, it is important to note that the ESG integration strategy focuses on the financial materiality of ESG factors to improve the risk management of a portfolio strategy, while an SDG-aligned listed equity strategy looks for company characteristics that are potentially impact-generating (Coqueret, 2022); it thus directs investors to companies that are explicitly delivering SDGs outcomes through their activities. In summary, there are four stages for integrating SDGs into investor decision-making:

- Exploration of green SDG opportunities and assets by optimizing financial returns;

---

[2] IRIS+ provides generally accepted core metrics sets aligned to common impact themes and Sustainable Development Goals. For further details: https://iris.thegiin.org/metrics/.

[3] GRI (Global Reporting Initiative) is an independent, international organization that helps businesses and investors in their impact reporting activities by providing the world's most widely used standards for sustainability reporting. For details: https://www.globalreporting.org/.

- Evaluation of exposures and related risks;
- Setting the goals aligned with the company's strategy, and integration into a portfolio;
- Measurement that exceeds a rating approach.

The risk assessment considers environmental and climate-related risks within the broader risk assessment framework of the entire portfolio, and at the various stages of the investment's lifecycle. Also, the measurement should routinely assess the environmental performance by considering quantifiable environmental outcomes of investments, and compare them against a set of green SDG impact standards.

The growth in equity products that seek to address the development and financing solutions for achieving green SDG outcomes has centered on exchange-traded funds (ETFs) and mutual funds, and less prominently by private equity solutions (Dovie, 2019; Miralles-Quirós et al., 2019). The main green-SDG ETF products and SDG mutual funds are reported in Table 4.3.

In conclusion, the SDGs provide new opportunities for the green finance sector to enhance its current green investing approach with SDG-aligned analysis, which helps to improve the sustainability risk and opportunity assessments of their strategies. Through the accurate assessment of these issues, financial actors will be able to identify a new environmental value from SDG-aligned capital allocation. In this context, data and index providers will have a fundamental role in supplying tools

Table 4.3 The SDGs in green equity markets: some examples

Equity product structure	Main exemplar products
SDG-themed ETF	iShares MSCI Global Impact ETF
	Impact Shares Sustainable Development Goals Global Equity ETF
	Serenity Shares US Core Impact Index
SDG Mutual funds	Cornerstone Capital Access Impact Fund
	Federated Hermes SDG Engagement Equity Fund Institutional Shares
	AB Sustainable Global Thematic Fund
	RobecoSAM Global SDG Equities
	BMO SDG Engagement Global Equity Fund

Source Author's elaboration

needed by green investors, in order to capitalize on innovative SDG strategies for assessing and measuring the total green value achieved by their investment.

4.7 Conclusion

The chapter highlights how the green investment movement has evolved in recent years, moving from simple exclusion to ESG integration with investment decision-making. More recently, the green investment framework with an ESG emphasis has focused on companies whose activities contribute positively to green Sustainable Development Goals. They also provide a common language for adding environmental impact issues to the value creation chain of green investments.

Within this context, while selecting SDG investment opportunities among isolated companies or green projects is costly, due to the efforts required to track and measure the environmental outcomes, the sector of green-themed investing based on SDGs shows more rapid growth potential for sectors such as green agriculture, renewable energies, and climate-change-impactful green tech solutions. Indeed, a number of SDG-(green)themed equity ETFs and mutual funds have emerged. However, tools or databases for tracking such investments in the SDGs are still lacking, and report frameworks for SDG KPIs are still limited for investment purposes. Thus, the key challenge for green investors is to make green SDGs investable through alignment models that exceed the ESG rating by enhancing green investments' accountability for achieving SDGs.

Furthermore, this emerging trend has interesting implications for the advancement of the green finance market, for different reasons. First, integrating environmental sustainability via the SDGs is just one trend shaping the climate/environment conversation regarding funds. In addition, this approach provides an additional layer of transparency, for investors to better assess the *greenness* of their portfolio companies. Finally, if the impact related to an investment's SDG target is only realized years later, then, regardless of the availability of environmental impact data, investors must be willing to exercise less patience. Thus, SDG investment that addresses green themes seems better able to attract the *patience-averse* (but environmentally oriented) mainstream investors.

REFERENCES

Arora, N. K., & Mishra, I. (2019). United Nations Sustainable Development Goals 2030 and environmental sustainability: Race against time. *Environmental Sustainability, 2*(4), 339–342.

Avramov, D., Cheng, S., Lioui, A., & Tarelli, A. (2021). Sustainable investing with ESG rating uncertainty. *Journal of Financial Economics*.

Barman, E. (2018). Doing well by doing good: A comparative analysis of ESG standards for responsible investment. In S. Dorobantu, R. V. Aguilera, J. Luo, & F. J. Milliken (Eds.), *Sustainability, stakeholder governance, and corporate social responsibility*. Emerald Publishing Limited.

Bender, J., Bridges, T. A., He, C., Lester, A., & Sun, X. (2018). A blueprint for integrating ESG into equity portfolios. *The Journal of Investment Management, 16*(1).

Bender, J., Sun, X., & Wang, T. (2017). Thematic indexing, meet smart beta! Merging ESG into factor portfolios. *The Journal of Index Investing, 8*(3), 89–101.

Berg, F., Koelbel, J. F., & Rigobon, R. (2019). *Aggregate confusion: The divergence of ESG ratings* (pp. 1–42). MIT Sloan School of Management.

Billio, M., Costola, M., Hristova, I., Latino, C., & Pelizzon, L. (2021). Inside the ESG Ratings:(Dis) agreement and performance. *Corporate Social Responsibility and Environmental Management, 28*(5), 1426–1445.

Bocken, N., Boons, F., & Baldassarre, B. (2019). Sustainable business model experimentation by understanding ecologies of business models. *Journal of Cleaner Production, 208*, 1498–1512.

Boffo, R., & Patalano, R. (2020). *ESG investing: Practices, progress and challenges*. OECD. www.oecd.org/finance/ESG-Investing-Practices-Progress-and-Challenges.pdf

Broadstock, D. C., Chan, K., Cheng, L. T., & Wang, X. (2021). The role of ESG performance during times of financial crisis: Evidence from COVID-19 in China. *Finance Research Letters, 38*, 101716.

Brooks, C., & Oikonomou, I. (2018). The effects of environmental, social and governance disclosures and performance on firm value: A review of the literature in accounting and finance. *The British Accounting Review, 50*(1), 1–15.

Chan, Y., Hogan, K., Schwaiger, K., & Ang, A. (2020). ESG in factors. *The Journal of Impact and ESG Investing, 1*(1), 26–45.

Chevrollier, N., Zhang, J., van Leeuwen, T., & Nijhof, A. (2020). The predictive value of strategic orientation for ESG performance over time. *Corporate Governance: The International Journal of Business in Society, 20*(1), 123–142.

Chew, E., & Childe, M. (2019). How asset managers can better align public markets investing with the SDGs. In *Sustainable Development Goals:*

Harnessing Business to Achieve the SDGs through Finance, Technology, and Law Reform (pp. 143–166). Wiley.

Chiappini, H., Vento, G., & De Palma, L. (2021). The impact of covid-19 lockdowns on sustainable indexes. *Sustainability, 13*(4), 1846.

Chouaibi, S., Chouaibi, J., & Rossi, M. (2021). ESG and corporate financial performance: The mediating role of green innovation: UK common law versus Germany civil law. *EuroMed Journal of Business, 17*(1), 46–71.

Coqueret, G. (2022). *Perspectives in sustainable equity investing*. CRC Press.

Cort, T., & Esty, D. (2020). ESG Standards: Looming challenges and pathways forward. *Organization & Environment, 33*(4), 491–510.

De Franco, C., Nicolle, J., & Tran, L. A. (2021). Sustainable investing: ESG versus SDG. *The Journal of Impact and ESG Investing, 1*(4), 45–62.

Deng, X., & Cheng, X. (2019). Can ESG indices improve the enterprises' stock market performance?—An empirical study from China. *Sustainability, 11*(17), 4765.

Dimson, E., Marsh, P., & Staunton, M. (2020). Divergent ESG ratings. *The Journal of Portfolio Management, 47*(1), 75–87.

Dorfleitner, G., Kreuzer, C., & Sparrer, C. (2020). ESG controversies and controversial ESG: About silent saints and small sinners. *Journal of Asset Management, 21*(5), 393–412.

Doumbia, D., & Lauridsen, M. L. (2019). Closing the SDG financing gap: Trends and data. *EMCompass, 73*, 1–8.

Dovie, D. B. K. (2019). Case for equity between Paris Climate agreement's Co-benefits and adaptation. *Science of the Total Environment, 656*, 732–739.

ElAlfy, A., Palaschuk, N., El-Bassiouny, D., Wilson, J., & Weber, O. (2020). Scoping the evolution of corporate social responsibility (CSR) research in the sustainable development goals (SDGs) era. *Sustainability, 12*(14), 5544.

Elder, M., & Olsen, S. H. (2019). The design of environmental priorities in the SDGs. *Global Policy, 10*, 70–82.

Elsenhuber, U., & Skenderasi, A. (2020). ESG investing: The role of public investors in sustainable investing. *Evolving Practices in Public Investment Management*, 45.

Folqué, M., Escrig-Olmedo, E., & Corzo Santamaría, T. (2021). Sustainable development and financial system: Integrating ESG risks through sustainable investment strategies in a climate change context. *Sustainable Development, 29*(5), 876–890.

Friede, G., Busch, T., & Bassen, A. (2015). ESG and financial performance: Aggregated evidence from more than 2000 empirical studies. *Journal of Sustainable Finance & Investment, 5*(4), 210–233.

Gibson Brandon, R., Krueger, P., & Schmidt, P. S. (2021). ESG rating disagreement and stock returns. *Financial Analysts Journal, 77*(4), 104–127.

Giese, G., Lee, L. E., Melas, D., Nagy, Z., & Nishikawa, L. (2019). Foundations of ESG investing: How ESG affects equity valuation, risk, and performance. *The Journal of Portfolio Management, 45*(5), 69–83.

Global Sustainable Investment Alliance. (2021). *Global Sustainable Investment Review 2020*. http://www.gsi-alliance.org/wp-content/uploads/2021/08/GSIR-20201.pdf

Gyönyörová, L., Stachoň, M., & Stašek, D. (2021). ESG ratings: Relevant information or misleading clue? Evidence from the S&P Global 1200. *Journal of Sustainable Finance & Investment*, 1–35.

Halbritter, G., & Dorfleitner, G. (2015). The wages of social responsibility—Where are they? A critical review of ESG investing. *Review of Financial Economics, 26*, 25–35.

Impact Management Project. (2022). *Impact management norms*. https://impactmanagementproject.com/impact-management/impact-management-norms/

Jacob, A., & Wilkens, M. (2021, June 26). *What drives sustainable indices? A framework for analyzing the sustainable index landscape.*

Johnson, L., Sachs, L., & Lobel, N. (2019). Aligning international investment agreements with the Sustainable Development Goals. *Columbia Journal of Transnational Law, 58*, 58.

Johnsson, F., Karlsson, I., Rootzén, J., Ahlbäck, A., & Gustavsson, M. (2020). The framing of a sustainable development goals assessment in decarbonizing the construction industry—Avoiding "Greenwashing." *Renewable and Sustainable Energy Reviews, 131*, 110029.

Kölbel, J. F., Heeb, F., Paetzold, F., & Busch, T. (2020). Can sustainable investing save the world? Reviewing the mechanisms of investor impact. *Organization & Environment, 33*(4), 554–574.

Kotsantonis, S., Pinney, C., & Serafeim, G. (2016). ESG integration in investment management: Myths and realities. *Journal of Applied Corporate Finance, 28*(2), 10–16.

Krosinsky, C., & Robins, N. (Eds.). (2012). *Sustainable investing: The art of long-term performance*. Routledge.

Lassala, C., Orero-Blat, M., & Ribeiro-Navarrete, S. (2021). The financial performance of listed companies in pursuit of the Sustainable Development Goals (SDG). *Economic Research-Ekonomska Istraživanja, 34*(1), 427–449.

Lee, J. W. (2020). Green finance and sustainable development goals: The case of China. *The Journal of Asian Finance, Economics, and Business, 7*(7), 577–586.

Liang, H., Fernandez, D., & Larsen, M. (2022). 23. Impact assessment and measurement with sustainable development goals. *Handbook on the Business of Sustainability: The Organization, Implementation, and Practice of Sustainable Growth*, 424.

Linnér, B. O., & Selin, H. (2013). The United Nations conference on sustainable development: Forty years in the making. *Environment and Planning C: Government and Policy, 31*(6), 971–987.

Madaleno, M., Dogan, E., & Taskin, D. (2022). A step forward on sustainability: The nexus of environmental responsibility, green technology, clean energy and green finance. *Energy Economics, 109*, 105945.

Miralles-Quirós, J. L., Miralles-Quirós, M. M., & Nogueira, J. M. (2019). Diversification benefits of using exchange-traded funds in compliance to the sustainable development goals. *Business Strategy and the Environment, 28*(1), 244–255.

Muhmad, S. N., & Muhamad, R. (2021). Sustainable business practices and financial performance during pre-and post-SDG adoption periods: A systematic review. *Journal of Sustainable Finance & Investment, 11*(4), 291–309.

Nykvist, B., & Maltais, A. (2022). Too risky—The role of finance as a driver of sustainability transitions. *Environmental Innovation and Societal Transitions, 42*, 219–231.

Pisani, F., & Russo, G. (2021). Sustainable finance and COVID-19: The reaction of ESG funds to the 2020 crisis. *Sustainability, 13*(23), 13253.

Porter, M., Serafeim, G., & Kramer, M. (2019). Where ESG fails. *Institutional Investor, 16*(2).

Rajesh, R., & Rajendran, C. (2020). Relating environmental, social, and governance scores and sustainability performances of firms: An empirical analysis. *Business Strategy and the Environment, 29*(3), 1247–1267.

Reiser, D. B., & Tucker, A. (2019). Buyer beware: Variation and opacity in ESG and ESG index funds. *Cardozo Law Review, 41*, 1921.

Scheyvens, R., Banks, G., & Hughes, E. (2016). The private sector and the SDGs: The need to move beyond 'business as usual'. *Sustainable Development, 24*(6), 371–382.

Trabacchi, C., & Buchner, B. (2019). Unlocking global investments for SDGs and tackling climate change. In *Achieving the sustainable development goals through sustainable food systems* (pp. 157–170). Springer.

United Nations. (2016). *Transforming our world: The 2030 Agenda for Sustainable Development.* https://stg-wedocs.unep.org/bitstream/handle/20.500.11822/11125/unepswiosm1inf7sdg.pdf?sequence=1

United Nations. (2018). *Unlocking SDG financing: Good practices from early adopters.* https://unsdg.un.org/sites/default/files/Unlocking-SDG-Financing-Good-Practices-Early-Adopters.pdf

Vojtko, R., & Padysak, M. (2019). *Quant's look on ESG investing strategies.* Available at SSRN 3504767.

Walter, I. (2020). Sense and nonsense in ESG ratings. *Journal of Law, Finance and Accounting, 5*(2), 307–336.

Whelan, T., Atz, U., Van Holt, T., & Clark, C. (2021). ESG and financial performance. *Uncovering the Relationship by Aggregating Evidence from, 1,* 2015–2020.

Zhan, J. X., & Santos-Paulino, A. U. (2021). Investing in the sustainable development goals: Mobilization, channeling, and impact. *Journal of International Business Policy, 4*(1), 166–183.

Zhao, C., Guo, Y., Yuan, J., Wu, M., Li, D., Zhou, Y., & Kang, J. (2018). ESG and corporate financial performance: Empirical evidence from China's listed power generation companies. *Sustainability, 10*(8), 2607.

CHAPTER 5

Beyond Greenwashing: An Overview of Possible Remedies

5.1 Introduction

In the last decade, the rapid growth of green asset allocations among financial institutions has caused a corresponding increase in regulatory initiatives designed to prevent or mitigate the risk of *greenwashing*. This term usually refers to the gap between symbolic and substantive actions, in line with the stated sustainability approach (Siano et al., 2017). The diffusion of this practice has been explored according to its conceptual dimensions, as well as by identifying possible reasons for its emergence (Yang et al., 2020). An expected increase in financial and/or reputational performance, or the pressure concerning sustainability practices from different classes of stakeholders, have encouraged companies to address sustainability requirements but there are no concrete measures for testing such intentionality (Ferrón-Vílchez et al., 2021).

Greenwashing is a relatively new term within financial studies, and is rooted in other corporate sectors of research. In more detail, evidence of this unethical practice was initially sought by researchers in marketing, management, and Corporate Social Responsibility (CSR) studies (Alves, 2009; Bowen & Aragon-Correa, 2014; Parguel et al., 2015). However, in recent years, the greenwashing debate has also expanded into finance research. Indeed, the pressure to report sustainability information has also influenced financial actors; they have included greenwashing issues in their strategies, in order to mitigate reputational risk, and to fulfill their responsibilities to external stakeholders by reporting reliable and trustworthy information on their green performance (Bowers et al., 2020). Furthermore, the financial industry's growing attention to greenwashing themes also reflects the fact that the variety of internationally accepted standards and evaluations of the *greenness* of an investment could hinder the mobilization of private capitals into the green finance sector (In & Schumacher, 2021). In this evolving arena, the European Union is a particularly active and innovative regulatory actor; it has introduced a series of actions that provide a *mandatory* framework for investors, to enhance their investments' alignment with environmentally beneficial activities.

This chapter provides an overview of greenwashing in the financial sector and highlights possible remedies. First, the conceptualization of greenwashing in the investment sector is still rather ambiguous; therefore, the chapter adopts the European Union (2021) definition of greenwashing, as a practice of gaining an unfair competitive advantage by marketing a financial product as environmentally labeled when environmental standards have been not met. Thus, after briefly discussing the background of greenwashing's origin and concepts (Sect. 5.2), the chapter provides a review of global regulatory actions (Sect. 5.3), and then focuses on the role of taxonomies, ratings, and standards in this context (Sect. 5.4). Next, the study describes greenwashing signals in the financial sector, by providing a detailed classification (Sect. 5.5). Finally, a range of possible greenwashing remedies are discussed in Sect. 5.6, while 5.7 concludes.

5.2 The Conceptual Background of Greenwashing

The concept of greenwashing emerged in the 1990s, after a book by Greer and Bruno (1996) analyzed the marketing attitude of 20 big corporations toward their environmental intentions and behaviors. This book represents the first effort to introduce the complex concept of irresponsible behaviors, during the era of "corporate environmentalism" (Bowen, 2014; Phillips, 2019). Indeed, the first theoretical findings on this topic were obtained from marketing studies in the last century's final decade, when companies started to promote their environmentalism to attract a growing green-aware segment, by recognizing environmental concerns as a source of competitive advantage. However, when pursuing this strategy, companies often used green claims that were vague, and at times even false, with corporate reputational consequences (Laufer, 2003). The failure to fulfill communication promises, and the implications for corporate performance, formed the basis of the first conceptualizations of greenwashing, in terms of manipulation of public knowledge (Beder, 1998); the concept also drew on empirical investigations of green claims and the verification of related products (see, for instance, Terrachoice, 2007).

In the 2000s, the conceptualization of greenwashing was expanded in studies concerning Corporate Social Responsibility (CSR) (Babcock, 2010; Cherry & Sneirson, 2010). Specifically, organizations' strategy to maintain a good reputation among their stakeholders by engaging in CSR activities frequently resulted in a greenwashing attitude, which was not supported by actual behavior. The drivers of such irresponsible behavior were identified at external, organizational, and individual levels (Delmas & Burbano, 2011). In detail, organizational-level drivers are typical in large corporations which face higher levels of pressure from investors than smaller firms. Individual drivers refer to the impact of greenwashing issues for managers in their decision-making process; and external drivers are essentially represented by all forms of regulations imposed to prevent greenwashing.

Greenwashing: Key Features and Conceptualization

Studies concerning greenwashing practices, though they originated in different research fields (Akturan, 2018; Furlow, 2010; Lane, 2013; Laufer, 2003; Mitchell & Ramey, 2011; Vos, 2009; Watson, 2017; Wu et al., 2020), have converged toward a common understanding of the greenwashing concept. All refer to this phenomenon as a gap between symbolic actions, often expressed in the form of communication, and substantive actions that are not in line with an environmentally friendly approach. As anticipated, this disconnection between the two types of actions was initially explored by corporate communication and CSR studies; they identified greenwashing as a related degenerative phenomenon that also implied market manipulation or customer fraud (Balluchi et al., 2020; Gatti et al., 2019). Briefly, the fundamental idea of greenwashing can be expressed in the form of "talking and not doing" (Cho et al., 2015), in the majority of cases. More rarely, it has been associated with manipulative and unethical, or even illegal, actions (de Freitas Netto et al., 2020).

Greenwashing has been identified in different forms; for instance, focusing on the prevailing idea of symbolic (and manipulative) communication (for instance, Vollero et al., 2016), or the disconnection between communication and organizational actions (Berrone et al., 2017). Such actions have been identified with the terms of *decoupling* and *attention deflection* (Siano et al., 2017). First, the decoupling form of greenwashing reflects a disconnection between the organization's structure and activities, and its green claims; or in some cases, false statements resulting from management practices to gain social legitimacy (Marquis et al., 2016; Terrachoice, 2007). In such a form, corporations claim to fulfill stakeholders' expectations, but without real changes in organizational policies. The second type of greenwashing concerns celebrating a green corporate narrative, to deflect stakeholders' attention from unethical business practices (Lyon & Montgomery, 2015). The deflection activities can be pursued mainly by selective disclosure (for instance, by voluntarily providing an incomplete comparison), or by misleading reporting (Seele & Gatti, 2017).

More recently, the taxonomy of greenwashing was extended beyond symbolic irresponsible actions, to include substantial irresponsible actions (Siano et al., 2017). In other words, this form of greenwashing refers to deceptive communication to generate manipulative business practices. Corporate sustainability is asserted in a set of statements that satisfy shareholders' need for sustainability reputational returns; Perez (2015) examined this form of greenwashing in the light of corporate legitimacy theory.

To summarize, the forms of greenwashing identified from marketing and CSR studies may be classified as three main types: decoupling, attention deflection, and deceptive conduct as a sin of an unethical corporate legitimacy strategy. These can be pursued through either symbolic (communications) or executional actions; hence, such forms may focus on a product or service level (de Freitas Netto et al., 2020), or extend to firm level.

5.3 A Review of Global Greenwashing Regulatory Initiatives

Non-market external drivers of greenwashing in the financial sector are essential factors in its actual impact. Indeed, several regulatory efforts have been made worldwide to avoid this unethical phenomenon; the European Union has played a leading role in green investments since 2018, by adopting a number of initiatives to enhance transparency in the financial sector. An important milestone in this strategy was the issuing of the EU Taxonomy, a uniform classification system of sustainable economic activities that favors green investments and mobilizes private capitals for green activities (Nedopil Wang et al., 2020). However, the Taxonomy represents only one of the mandatory pieces of the EU's sustainability strategy for greening the European financial sector.

Regarding the topic of this chapter, it is especially important to note that:

- The Regulation 2019/2088 of November 27, 2019, on **sustainability-related disclosures in the financial services sector** (SFDR), requires financial firms to make detailed disclosures of their products and services' alignment with environmental or sustainable objectives, in order to understand what they deliver in terms of environmental benefits. It excludes bank accounts and loans, as well as structured products and derivatives.
- The **Taxonomy Regulation** was introduced with the EU Regulation 2020/852 on June 18, 2020, to establish a framework to facilitate sustainable investment, and the related amending Regulation 2019/2088. It is a classification framework for assessing how activities align with what is green and what is not. The EU Taxonomy is based on the following main environmental objectives:

    - Climate change mitigation;
    - Climate change adaptation;
    - The sustainable use and protection of water and marine resources;
    - The transition to a circular economy;
    - The protection and restoration of biodiversity and ecosystems.

      The Taxonomy alignment of a company requires the disclosure of performance regarding the mentioned objectives, and the statement that none of the six are significantly harmed. Thus, the purpose of the EU Taxonomy is to create a more transparent structure for green investments, and to encourage green capital flows. The central point is that the Taxonomy requires disclosure of the product or service's alignment with what is considered green under the Taxonomy; thus, its function is considered supplementary to the Non-Financial

Reporting Directive. The Taxonomy is planned to come into force in 2022.

- The **European Green Bond (EUGB) regulation** was announced in the European Green Deal Investment Plan of January 14, 2020. In July 2021, the European Commission explored the possibility of a legislative initiative for a European Green Bond Standard through a public consultation on the renewed sustainable finance strategy, by issuing a proposal for Regulation on European Green Bonds. The main goals are: (i) to provide a uniform green bond standard within the EU, useful for issuers and investors of green bonds; (ii) to identify and trust high-quality green bonds while reducing the risk of greenwashing; and (iii) to stimulate the issue of green bonds and thereby attract private capital flows to green investments. It is notable that the bond standards require issuers to align the bond proceeds with the Taxonomy.
- The **Non-Financial Reporting Directive** (NFRD), 2014/95/EU (which is directed to large companies with over 500 employees, to disclose their information related to the sustainable transition) was reformed into the Corporate Sustainability Reporting Directive. The European Commission proposed the review on April 21, 2021, in its ambitious new package of "sustainable finance" regulation proposals, with the adoption of a new delegated act. With the reform adoption, this information disclosure framework will be extended to cover how and to what extent the corporate activities align with those included in the EU Taxonomy.

Outside the EU regulatory framework, regulatory bodies in the United Kingdom have also highlighted the increased greenwashing risk. In September 2021, the Competition and Market Authority published the Green Claims Code, a guidance to ensure that green claims meet regulatory expectations, by detailing six principles regarding the accuracy, clarity, information selection, truthful comparisons, and the substantial use of relevant evidence.

Two months before, the Financial Conduct Authority (FCA) had published a letter of recommendation for sustainable investment fund managers, about the delivery and disclosure of environmental, social, and governance (ESG) and green investment funds. The letter emphasized the need for fund managers to clearly describe their investment strategy, and to make substantiated assertions of their environmental and social goals. The recommendation was derived from an analysis of the ESG or sustainability focus of a number of investment funds that were largely identified as poorly drafted and disconnected from the FCA's expectations. The letter included guidelines which set out the rules applicable to authorized funds, to align them with ESG and sustainability standards. The guidelines comprise four principles:

- An overarching principle for ESG/sustainable investment funds to consistently reflect the stated sustainability mission in their name, stated objectives strategy, and documents;
- Consistency with the fund's sustainability mission in the design of the fund, and the relevant disclosure in their documents;
- Consistency with the fund's sustainability mission in the delivery of funds, by providing guidance on resources, data and analytical tools, and holdings, in pursuit of the stated sustainability mission;
- Consistency with the fund's sustainability mission in the ongoing periodic disclosures on funds, to help investors make investment decisions.

Finally, the FCA proposed a set of Climate-Related Financial Disclosure rules for asset managers, to come into force in 2022, with the aim of protecting consumers from buying non-environmentally friendly products.

On a policy level, the Government of the United Kingdom announced:

- The proposal of "Sustainability Disclosure Requirements" that requires financial actors to report their impact on climate and the environment, and related risks facing their business;
- The creation of a new sustainable investment label for compliant products;
- The release of a policy document[1] regarding the long-term ambition to green the financial system;
- The creation of a new expert group to help tackle greenwashing, which will oversee the UK Government's delivery of a "Green Taxonomy," intended as a common framework to define investments as environmentally sustainable.

---

[1] "Greening Finance: A Roadmap to Sustainable Investing," published on October 18, 2021. Available at: https://www.gov.uk/government/publications/greening-finance-a-roadmap-to-sustainable-investing.

On a global level, it is important to highlight the greenwashing initiatives of the US Securities and Exchange Commission (SEC), of the Hong Kong Securities and Futures Commission (SFC), and of the International Organization of Securities Commissions (IOSCO). Specifically, in March 2021, the US SEC announced the creation of a climate and ESG Taskforce, which would identify the misleading ESG practices in the context of investor disclosures, by focusing mainly on greenwashing misconduct related to the existing SEC disclosure guidance regarding climate change. In Hong Kong, in June 2021 the SFC announced guidance that, from January 2022, will help all ESG funds and climate-focused fund products to periodically disclose how ESG factors are incorporated and reported in their activities and portfolios. Finally, in June 2021, IOSCO published a Consultation Report[2] proposing that securities regulators develop frameworks for asset managers regarding sustainability-related risks and opportunities, by outlining the types of greenwashing at the asset manager level.

5.4 The Role of Green Taxonomies, Ratings, and Standards

This section provides an overview of the three main tools, i.e., taxonomies, ratings, and standards, adopted both to screen green finance targets and to disclose green investments. Indeed, there is broad consensus that such tools are considered as key elements supporting green finance in the transition to a more sustainable economy. In this context, it is important to highlight how such tools should be designed to achieve this goal in the most effective way.

Green Taxonomies: An Overview and an Analysis of Key Dimensions

A taxonomy for green finance is a set of criteria that assists in evaluating whether and to what extent a financial asset is aligned with supporting given environmental targets (Dusík & Bond, 2022). In order to achieve an objective taxonomy, it is useful to obtain a classification of assets by their degree of support for given environmental benefits (Tripathy et al.,

---

[2] The Report, entitled "Report on Sustainability-related Issuer Disclosures" is available on-line at: https://www.iosco.org/library/pubdocs/pdf/IOSCOPD678.pdf.

2020). Such assessment is based on information that investors, as well as other stakeholders, derive from the taxonomy (Migliorelli & Dessertine, 2019; Pacces, 2021). In this sense, taxonomies have essentially two functions:

- Assisting investors in their decision-making to join the pre-established strategy, to align their assets with the broader policy or regulatory environment-related targets;
- Provide investors and other stakeholders with the information necessary to implement disclosure requirements, regarding their asset allocation strategy's compliance with environmental objectives.

It is important to note that a taxonomy classifies a single asset and, therefore, does not provide information related to the interrelation with other assets. Such information could serve as a risk management tool; for example, to understand the degree of exposure to climate risk, or to the boarding of a specific planetary boundaries risk (Butz et al., 2018). Thus, the ultimate goal of a taxonomy is to assure investors that their investments are effectively contributing to defined environmental goals.

In the analysis of such a measure, it is possible to observe different classes of taxonomies in green finance. Indeed, beyond the European Union's taxonomy, as previously described in Sect. 5.3, China has also issued an official green taxonomy containing a catalogue of Green Bond Endorsed Projects delivered by the People's Bank of China (PBOC, 2021), the National Development and Reform Commission (NDRC), and the China Securities Regulatory Commission (CSRC) (Macaire & Naef, 2021). Furthermore, the market-based taxonomy of the Climate Bond Initiative (CBI) is also included among the most prominent and adopted green taxonomies (Vaze et al., 2019). In particular, taxonomies can be categorized according to their scope, target, or result.

Firstly, analyzing the taxonomies reveals whether their objectives are consistent with existing national or supranational standards and regulations; or, more broadly, with high-level policy goals. In this regard, the EU Taxonomy and CBI Taxonomy are strictly related to the net-zero emission target declared by the Paris Agreement of 2015. Furthermore, such objectives have multiple aspects, such as climate change mitigation, the protection of water and marine resources, transition to a circular economy, or pollution prevention. For this reason, the principle of *Do No*

Significant Harm is added, in order to alleviate possible conflicts among different objectives when screening for eligible activities (Jenkins, 2022). Multiple objectives are stated in the EU and Chinese taxonomies, while the CBI Taxonomy is essentially focused on climate mitigation targets.

Secondly, focusing on the target dimension, the most widely used taxonomies define sustainability from the perspective of the activity, rather than in terms of the entire entity (usually a corporation) undertaking the activity. Thus, the related metrics are generally activity-based. In some cases, they can be asset-based if the taxonomy targets assets on the entity's balance sheet, as in the case of the CBI Taxonomy.

Finally, regarding the intended output of taxonomies, it is primarily expressed in terms of ensuring data availability and the disclosure of financial data for different types of financial products. A second level of output from the adoption of taxonomies is related to facilitating the verification activity by third parties, even if a standardized verification process is still lacking. Among the verification models, the CBI seems to be the most advanced and structured of the main taxonomy frameworks. A third level of output relates to the extent of the taxonomies' screening of green activities; this can be divided into simplified clustering (brown vs green activities), or granular clustering of activities in different shades of eligibility. An overview of the use of taxonomies in green finance is presented in Table 5.1.

The Development of Green Ratings

The growing demand for green investments has increased the adoption of tools that indicate to what extent the investment target can be labeled as green-compliant or environmentally impactful. The use of ratings, largely identified as ESG ratings, includes different green finance instruments, ranging from green bonds and green loans, to stocks and mutual funds.

The development of ratings for green bonds has reached a mature stage within the sector. Furthermore, many credit-rating agencies have recently developed ESG scores for issuers, separate from the ratings of green bonds, which usually apply only to a specific product. It is important to note that ESG scoring systems have been introduced for all asset classes, including green insurance and sovereign green bonds (Li et al., 2021). ESG Risk Ratings can help investors measure the degree to which ESG issues are putting a company's enterprise value at risk. The higher the ESG Risk Rating, the higher the company's unmanaged

Table 5.1 The use of green taxonomies in the green finance sector

Key dimensions	Users	Green taxonomy list
• Alignment with priority environmental objectives → The taxonomy includes assets that support national/international environmental policy objectives • Balancing co-dependency → When green taxonomy incorporates multiple green targets (e.g. EU Taxonomy) an eligible activity addressing one of the environmental targets must also do no significant harm to any of the others • Definition of the scope and level of granularity → Taxonomy should clearly indicate: the scope, by specifying if harmful activities—but with available pathways to zero emissions—are eligible or not; non-binary granularity should be indicated to distinguish different levels of activity's greenness (e.g. by traffic lights)	• Banks and Financial Institutions → For consistent structuring of green banking products and lending activities while reducing uncertainty and reputational risk • Financial Regulators → For the prevention of greenwashing and to facilitate environmental sustainability reporting and disclosure guidelines • Green Investors → For the design of green-aligned portfolios and disclose exposure to sustainable/green investments as required by regulators • Green Bond Issuers → For the identification of green eligible activities to be financed with the green bond issuing • Policymakers →	• The European Union's Taxonomy • Climate Bond Initiative Taxonomy • Mongolia's Green taxonomy • China's Green Bond Endorsed Projects Catalogue • Malaysian Climate Change and Principle-Based Taxonomy • ASEAN Taxonomy • Bangladesh Sustainable Finance Taxonomy • Indonesia Green taxonomy

(continued)

Table 5.1 (continued)

Key dimensions	Users	Green taxonomy list
• Availability of unit of measurement and data disclosure → Most widely used green taxonomies adopt activity-based metrics with thresholds (related to the predetermined environmental target) for eligibility; furthermore the disclosure of the taxonomy's alignment of the green financial product is needed to ensure the integrity, transparency, consistency, and comparability of green labels across financial markets	For the development of strategies to achieve national/international environmental development commitments and for bridging the funding gap in green areas of underinvestment	

Source Author's elaboration

ESG risk. In this sense, ESG-based ratings are a practical way to support investment decisions. Moreover, ESG-based ratings are also effective in dealing with increasingly stringent information disclosure requirements.

However, the main challenge appears to be the fragmentation of ratings models. Specifically, the differences between green ratings depend on how investments that are considered green are reported in enterprises' information disclosures. As a result, different institutions use their own methods of calculating ESG-based ratings. For this reason, ESG information disclosure requirements are expected to become more stringent in future. Indeed, the increased assets of funds with an environmental, social, and governance (ESG) mandate registered in the last years has raised a question about the role of ESG green ratings in greenwashing practices (Gyönyörová et al., 2021). For instance, asset managers who claim to follow an ESG-led mandate but invest in companies with significant ESG-related risk profit from duping consumers who believe they are making an earth-friendly or socially conscious choice. Acknowledging increased investor demand for ESG information, the European and the US Regulatory Market Authorities (respectively, the European Securities and Markets, and the Securities and Exchange Commission) are converging toward the conclusion that regulatory and supervisory efforts are needed for ESG ratings. The same improvements are called also for all other kinds of green ratings on whether investments are sustainable and climate-friendly, in order to avoid investors being deceived by greenwashing (Lee & Suh, 2022; Zhang, 2022).

Voluntary Market Standards and Self-Regulatory Initiatives

Over the past decade, there has been considerable growth in self-regulatory initiatives which have attempted to establish voluntary market standards within the green finance industries (Troeger & Steuer, 2021). They can be grouped into three main clusters:

- Sustainable and responsible business practices;
- Disclosure-related initiatives;
- Sustainable investing initiatives directed at company disclosure and metrics, business behavior and standards.

In the first group, the Global Compact[3] is the framework most widely adopted by companies committed to sustainability and responsible business practices; while in the reporting cluster, the most commonly used are the Sustainability Accounting Standards Board (SASB), the Task Force on Climate-Related Financial Disclosures (TCFD), the Global Reporting Initiative (GRI), and the World Business Council for Sustainable Development (WBCSD) (Shoaf et al., 2018). Among investment frameworks, the Principles for Responsible Investment[4] is the most adopted, which helps responsible investors to incorporate ESG factors into their investment and ownership decisions. However, a "one-size-fits-all" green investment standard is currently difficult to develop, considering that local contexts influence the relevance of green targets, and there may be different geographical and economic conditions within individual adopting countries.

Transparent measurement and disclosure of environmental performance are now considered a fundamental part of investment decision in the green finance market. With the rise of sustainability reporting demand and the complexity of the accounting and reporting standards and measures, the reliability of such information became a major issue among relevant stakeholders. For these reasons, the disclosure of green metrics and information according to these standards are encouraged. Yet, several issues remain, considering the complexity surrounding these, often voluntary, tools. For example, the compliance with such frameworks does not exclude ambiguity and selectivity from the materiality analyses. Furthermore, especially sustainability accounting frameworks lack the infrastructure of finance reporting. In other terms, they lack the same weight and rigor adopted for financial reporting, which are mandatory as well as external audited.

---

[3] The United Nations Global Compact is a strategic initiative that supports global companies that are committed to responsible business practices in the areas of human rights, labor, the environment, and corruption. The framework does not provide a code of conduct for corporations with monitoring or verification procedures; rather, it relies on public accountability, transparency, and enlightened self-interest to fulfill and disclose their aims. For further details: https://www.unglobalcompact.org/.

[4] The Principles were developed by an international group of institutional investors and were convened by the United Nations Secretary-General. The six principles encourage investors to use responsible investment to enhance returns and better manage risks; they provide a set of possible actions for incorporating ESG issues into investment practice. For more details: https://www.unpri.org/about-us/about-the-pri.

5.5 A Classification of Greenwashing Signals in the Green Financial Sector

The phenomenon of greenwashing in the financial sector assumes different aspects. For instance, the increasing pressure on sustainability reporting among financial actors has raised a series of questions about the quality of information disclosed (Khan et al., 2020). The recent policy initiatives, in both developed and developing countries, have globally highlighted the importance of financial actors' true and reliable sustainability reporting practices, by pushing for green transparency, which may also open markets to the transition in the future. The banking sector, in particular, is the financial sector most interested in increasing non-financial reporting activities (Loew et al., 2019). Indeed, asset managers, funds, and banks, represent the main targets of the above-described regulatory frameworks. In simplified terms, with such pressure on financial institutions to address environmental and social change, the risk of greenwashing has become a real and material issue; especially if such actors do not set transparent, ambitious, and measurable targets.

To mitigate the risk of greenwashing among financial markets, regulatory actors have mostly focused on the theme of defining what are green activities and what are not. However, the forms of greenwashing in the green financial sector may vary across products/services and actors. For instance, stakeholders may also pressurize non-financial companies for greener actions; this can encourage greenwashing, which results in adverse reactions from investors, as well as damage to the firm's image. Thus, a holistic map of greenwashing in green finance can be derived by combining (*i*) the literature that maps the empirical forms of greenwashing (de Freitas Netto et al., 2020; de Jong et al., 2020; Falcão et al., 2020; Jones, 2019; Lyon & Montgomery, 2015; Marciniak, 2009; Marquis et al., 2015, 2016; Parguel et al., 2011; Siano et al., 2017; Yang et al., 2020; Yu et al., 2020) (*ii*) the evolving regulatory framework in the green finance sector regarding this issue, as outlined above and, (*iii*) the green finance ecosystem.

Primarily, it is important to note that unlike greenwashing at the business level, this phenomenon in financial markets produces broader negative effects. Indeed, the role of green finance as "finance for transition" can be eroded by greenwashing effects, in terms of reputational crisis and shaking the market's confidence, with consequences that can translate into reduced capital availability for green projects.

The forms of greenwashing in green finance can be framed as:

- Actions related to green financial products or services;
- Actions related to corporations' green alignment;
- Actions related to green investors (asset managers and financial institutions overall).

Figure 5.1 summarizes the described scheme.

Regarding the greenwashing actions related to green financial products or services, unreliable certifications and labels are the main forms that can be observed. Indeed, certifications and labels can be an efficient way of reducing greenwashing, especially for green bonds. Untrustworthy certifications and labels for green financial products or services can strongly impact investors. Similar forms of unethical behaviors may also reveal a specific form of such practice: misleading by falsely asserting that a third party has been involved in the certification. Furthermore, this range of actions also includes misunderstanding created by vague claims disclosed in product-related reports.

The second group of actions can be associated with corporations, either investees or issuers of green financial products. Here, the main form of greenwashing practice is related to misleading actions of non-transparent

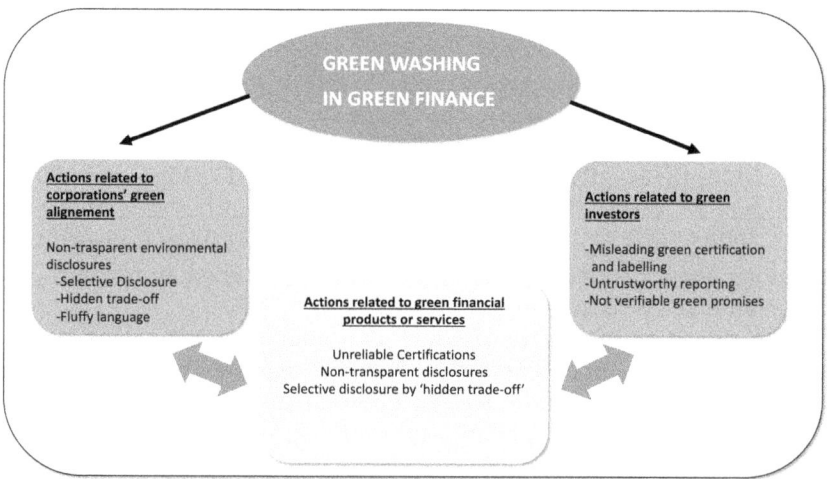

Fig. 5.1 Greenwashing in green finance (*Source* Author's elaboration)

environmental disclosures, in the form of "selective disclosure" through "hidden trade-offs" or "fluffy language." In the first case, the firm disproportionately focuses on one benign environmental performance by obscuring its overall negative environmental performance indicators. This creates a misleading optimistic narrative based on inaccurate claims, or on words with no clear meaning (fluffy language). Without a uniform set of criteria, such as a taxonomy of green activities, firms may reveal only parts of their information, which can obscure potential harms to environmental objectives.

Finally, the group of greenwashing actions related to green investors essentially concerns the adoption of misleading practices through green certifications and labels, non-transparent reporting, and manipulation of data on the extent of their green alignment. In the latter case, providing data without technical screening criteria according to a classification of sustainable activities can produce this type of greenwashing problem. In other words, this form provides environmental claims that cannot be verified without accessible supporting information.

An overview of cumulative cases of greenwashing in finance is reported in Table 5.2.

5.6 A Range of Possible Solutions to Prevent Greenwashing

As illustrated in the previous sections, proponents of green finance and financial regulators have reached a growing consensus that environmental factors need to be part of green investment decisions, but less agreement on how to put that into practice and define such non-financial parameters.

By attempting to provide a framework of possible remedies for greenwashing, it is possible to identify three main pillars for a greenwashing-solutions dashboard, which could be applied with the simultaneous and combined efforts of regulators, standard setters, and academia.

The first pillar relates to tools for identifying green investments; the second to incentives for increasing the participation of third-party verifiers; and the third concerns a roadmap for universal environmental information disclosure.

Remedy #1: Tools for the "Green" Alignment of Investments

Regarding the first pillar, it is interesting that in the described greenwashing regulation actions, the adoption of taxonomies or guidelines in

Table 5.2 Illustrative cases of greenwashing in green finance

Green finance instruments (Green, sustainability-linked and transition bonds)	Banking sector/Asset managers
• The identification of the environmental impact indicators in green bonds are usually assessed ex-ante without ex-post verification • The issuing of the green bond could be 'isolated' within a broader brown strategy of the issuer • When the product is related to the transition, the use of the term could serve to dilute future environmental benefits; furthermore the notion of transition is quite complex and requires issuer-level screening • In sustainability-linked bonds the choice of KPIs should privilege indicators not manipulable in their calculation methodologies • Sustainability-linked bonds could suffer washing practices deriving from a moral hazard associated to the major gains for investors when issuer fails the achievement of sustainability performances • The cost of achieving the environmental performances could be not proportionated to the issuance amount	• Inflation of the ESG credentials stated in annual reports • Green in name only—Investor firms fail to live up to ESG-related legal, regulatory or public obligations or commitments • Using a fund's green/ESG label without a minimum amount of green asset under management • Using a fund's green/ESG label without the pursuing of a sustainable investment strategy • Disclosures inconsistent with a green portfolio management practice • Existing discrepancies between public ESG-related proxy voting claims and internal voting policies • Definition of measurable green objectives (non-financial returns) presented in a non-proportionated way in a green product offer • Claiming an investment strategy to invest in companies contributing to "positive environmental impact" while investing predominantly in companies that report only low carbon emissions

Source Author's elaboration

identifying green activities is the most universally used tool for defining the boundaries between "green" and "brown" activities. Moreover, taxonomies represent one of the main tools that encourage the alignment of investments with green activities. In particular, there are found to be three main criteria for identifying an investment as green:

- The use of taxonomies, indicators, and ratings;
- The adoption of labels and certification schemes;
- The use of process criteria.

While the benefits deriving from the adoption of taxonomies, as deeply explored above, are related to their ability to allow stakeholders to easily identify whether certain items can be defined as green, such tools present some limitations due to the classification of green and brown activities.

This dichotomic classification may disadvantage those companies that, for instance, do not fall under one of the categories in the taxonomy, but have large potential for becoming greener.

On the other hand, the adoption of a rating system is useful for measuring the greenness of a firm, technology, or financial product, according to predefined criteria. In developing a green-related rating, three main steps are followed:

- Selecting indicators of sustainability exposure (or of sustainability performance) recognized in the sector in question;
- Gathering and assessing data;
- Calculating the rating by scoring and weighting data.

The scoring from sustainability-related rating providers is offered in different ways in the sector (Amariei, 2019); thus, the same target company may receive different and incomparable ratings, even with similar starting databases. This may be due to bias related to company size or to the type of industry sector typically covered by sustainability-related ratings providers. Thus, the use of green-related ratings by asset owners may differ significantly.

Differently from ratings, indices can provide investment managers and asset owners with a benchmark, usually for operating actively managed portfolios. The index classification systems may be based on identifying green companies by green sector or according to various taxonomies, as well as on exclusion criteria.

Finally, labels and certification schemes are also used to certify the greenness of an investor. Usually, such tools assess the eligibility of applicant financial actors using a green list of projects that are considered to contribute positively to the environment. Other cases require that the applicant financial actor has a certain percentage of assets invested in green-related activities identified by taxonomies. Such tools still have a limited use, and are adopted essentially for investment funds.

In terms of the described remedies to prevent greenwashing, the taxonomies, indicators, and ratings represent the main adopted tools. However, their effective application should take into consideration a series of critical issues. First, the plurality of taxonomies jeopardizes the identification of green activities within a geographical area. A reduction of this plurality could enhance cross-border and cross-market green investment, which would help to evaluate the green performance of financial firms, as well as the comparability of eligibility criteria. Furthermore, regarding ratings and indicators, the mentioned opacity and fragmentation in the rating methodology suggests the need to develop a transparent and standardized system for reporting performance on environmental or sustainability objectives. It is important to note that the data considered in a possible standardized rating methodology should be integrated by incorporating data on the impact over an investment's entire lifecycle, as well as open data and financial reporting (by financial authorities) of environmental crimes. Such solutions could represent further innovations and steps to ensure that the entire green financing value chain is free of environmental crimes.

Remedy #2: Increase Participation of Independent Evaluators

The growing attention toward non-financial disclosure and sustainability-related reporting represents a first step in the efforts to improve the transparency and trustworthiness of green financial actors and products. However, such measures could be less effective in preventing greenwashing practices if there are no uniform, internationally agreed standardized guidelines and overall mandatory rules. Efforts to encourage the participation of independent evaluators in "voluntary" green reporting and disclosure practices, as well as for the certification of proceeds when issuing green financial products, could increase the accountability and credibility of green financial investments, by ensuring that funds obtained through green financing are not allocated to assets with little or no environmental value.

Remedy #3: Develop Process Criteria for Untargeted Green Investments

A large proportion of green finance investments include those that are not directed to specific green projects or companies (i.e., untargeted green investments). In this typology of investments, capital funds companies are identified as successfully managing or being compliant with environmental (as well as social and governance) (ESG) factors and related risks. By adopting screening techniques, the best-performing organizations are identified as more environmentally friendly than others, especially those in listed and private equity investment, which belong to the broader segment recognized as sustainable and responsible investments (SRI). Also, capitals invested in green activities within corporate finance are usually provided for general purposes.

Considering this lack of correlation with a specific green purpose, the measure of an investment's environmental benefit is hard to determine. In such cases, the adoption of indicators and/or ratings, as outlined in the previous options, can provide guidance, mainly for green finance provided through the green use of proceeds bonds and green loans, or green project finance. For untargeted investments, alignment tools present difficulties in their adoption; while the process-oriented approach could be better suited to mitigating unintended greenwashing risks for this type of finance, by determining a set of requirements that focus on the process of green investments. For instance, process criteria can be adopted to harmonize the investment processes continuum, from ex-ante through to implementation, and also to ex-post financial decisions. Process criteria can determine, for example, the level of engagement with the investees, the information that should be disclosed about an investment or rating strategy, and the environmental impact of the investments. Overall, the expected environmental impact of untargeted green finance can be determined by assessing the impacts by adopting methodologies, indicators, and metrics existing within the impact investing industry, which is already based on specifically assessing and reporting on the impacts of an investment.

5.7 Conclusion

This chapter has explored the topic of greenwashing within green finance. After a review of the current state of this phenomenon, by providing the conceptual background and empirical insights, this study opens new perspectives on greenwashing in finance, by examining its different aspects within the financial sector, and particularly, among green investments. Furthermore, investigations at different levels have revealed its different forms within green finance segments, and in terms of the voluntary or regulatory tools to prevent this risk. Finally, an outline of possible remedies was presented.

In conclusion, based on the analysis performed in this chapter, it is possible to define three main results:

- The European Union market presents the most advanced framework for detecting and preventing greenwashing in green investments;

- Greenwashing remedies are overall concentrated over alignment tools, but in different ways along segments;
- Solutions for preventing greenwashing present a certain degree of opacity in the area of ESG-based investments, mainly in equity-based investments.

In order to achieve and preserve the market integrity of green investments, further steps need to be elaborated, agreed, and shared. The theme of green alignment appears inadequate to cover the complexity of the green finance market, as well as its geographical variations. The momentum of ESG investing has increased the risk of greenwashing, along with uncoordinated and unsupervised rating methodologies at an international level. In addition, taxonomies alone are not sufficient to identify what is green in the context of green finance, especially for untargeted green investments, such as in the case of ESG investing. Therefore, the progressive inclusion of the dimension of environmental impact was suggested by the consultation on a Renewed Sustainable Finance Strategy (European Commission, 2020), which included, among other concerns, an explicit focus on how different sustainable financial products impact the real economy. In other words, the policy focus on portfolios' exposure to certain green activities should include improvements enabling the recognition of the investor's impact in the real economy. This paradigm shift, from investing *with* to investing *for* green outcomes, could significantly reduce the greenwashing risk; but this will require an analysis that is more sophisticated (and costly) than observing the portfolio allocation or alignment.

References

Akturan, U. (2018). *How does greenwashing affect green branding equity and purchase intention? An empirical research.* Marketing Intelligence & Planning.

Alves, I. M. (2009). Green spin everywhere: How greenwashing reveals the limits of the CSR paradigm. *Journal of Global Change & Governance, 2*(1), 1–26.

Amariei, C. (2019). *Sustainability in practice: Ratings, research and proprietary models* (ECMI Policy Brief, 26).

Babcock, H. M. (2010). Corporate environmental social responsibility: Corporate greenwashing or a corporate culture game changer. *Fordham Environmental Law Review, 21*, 1–78.

Balluchi, F., Lazzini, A., & Torelli, R. (2020). CSR and greenwashing: A matter of perception in the search of legitimacy. In *Accounting, accountability and society* (pp. 151–166). Springer.

Beder, S. (1998). Manipulating public knowledge. *Metascience, 7*(1), 132–139.
Berrone, P., Fosfuri, A., & Gelabert, L. (2017). Does greenwashing pay off? Understanding the relationship between environmental actions and environmental legitimacy. *Journal of Business Ethics, 144*(2), 363–379.
Bowen, F. (2014). *After greenwashing: Symbolic corporate environmentalism and society*. Cambridge University Press.
Bowen, F., & Aragon-Correa, J. A. (2014). Greenwashing in corporate environmentalism research and practice: The importance of what we say and do. *Organization & Environment, 27*(2), 107–112.
Bowers, B., Boyd, N., & McGoun, E. (2020). Greenbacks, green banks, and greenwashing via LEED: Assessing banks' performance in sustainable construction. *Sustainability: The Journal of Record, 13*(5), 208–217.
Butz, C., Liechti, J., Bodin, J., & Cornell, S. E. (2018). Towards defining an environmental investment universe within planetary boundaries. *Sustainability Science, 13*(4), 1031–1044.
Cherry, M. A., & Sneirson, J. F. (2010). Beyond profit: Rethinking corporate social responsibility and greenwashing after the BP oil disaster. *Tulane Law Review, 85*, 983.
Cho, C. H., Laine, M., Roberts, R. W., & Rodrigue, M. (2015). Organized hypocrisy, organizational façades, and sustainability reporting. *Accounting, Organizations and Society, 40*, 78–94.
de Freitas Netto, S. V., Sobral, M. F. F., Ribeiro, A. R. B., & da Luz Soares, G. R. (2020). Concepts and forms of greenwashing: A systematic review. *Environmental Sciences Europe, 32*(1), 1–12.
de Jong, M. D., Huluba, G., & Beldad, A. D. (2020). Different shades of greenwashing: Consumers' reactions to environmental lies, half-lies, and organizations taking credit for following legal obligations. *Journal of Business and Technical Communication, 34*(1), 38–76.
Delmas, M. A., & Burbano, V. C. (2011). The drivers of greenwashing. *California Management Review, 54*(1), 64–87.
Dusík, J., & Bond, A. (2022). Environmental assessments and sustainable finance frameworks: Will the EU Taxonomy change the mindset over the contribution of EIA to sustainable development? *Impact Assessment and Project Appraisal*, 1–9.
European Commission. (2020). *Summary report of the stakeholder consultation on the renewed sustainable finance strategy*. https://ec.europa.eu/info/sites/default/files/business_economy_euro/banking_and_finance/documents/2020-sustainable-finance-strategy-summary-of-responses_en.pdf
European Commission. (2021). *Screening of websites for 'greenwashing': Half of green claims lack evidence*. Retrieved on 28 December 2021 from https://ec.europa.eu/commission/presscorner/detail/en/ip_21_269

Falcão, S. M. F., Bezerra, R. A. R., & da Luz, S. G. R. (2020). Concepts and forms of greenwashing: A systematic review. *Environmental Sciences Europe*, *32*(1).

Ferrón-Vílchez, V., Valero-Gil, J., & Suárez-Perales, I. (2021). How does greenwashing influence managers' decision-making? An experimental approach under stakeholder view. *Corporate Social Responsibility and Environmental Management*, *28*(2), 860–880.

Financial Conduct Authority (FCA). (2021). *Authorised ESG & Sustainable Investment Funds: Improving quality and clarity*. https://www.fca.org.uk/publication/correspondence/dear-chair-letter-authorised-esg-sustainable-investment-funds.pdf

Furlow, N. E. (2010). Greenwashing in the new millennium. *The Journal of Applied Business and Economics*, *10*(6), 22.

Gatti, L., Seele, P., & Rademacher, L. (2019). Grey zone in–greenwash out. A review of greenwashing research and implications for the voluntary-mandatory transition of CSR. *International Journal of Corporate Social Responsibility*, *4*(1), 1–15.

Greer, J., & Bruno, K. (1996). *Greenwash: The reality behind corporate environmentalism*. Third World Network.

Gyönyörová, L., Stachoň, M., & Stašek, D. (2021). ESG ratings: Relevant information or misleading clue? Evidence from the S&P Global 1200. *Journal of Sustainable Finance & Investment*, 1–35.

In, S. Y., & Schumacher, K. (2021). Carbonwashing: ESG Data Greenwashing in a Post-Paris World. In *Settling Climate Accounts* (pp. 39–58). Palgrave Macmillan.

Jenkins, B. (2022). Response to environmental assessments and sustainable finance frameworks. *Impact Assessment and Project Appraisal*, 1–5.

Jones, E. (2019). Rethinking greenwashing: Corporate discourse, unethical practice, and the unmet potential of ethical consumerism. *Sociological Perspectives*, *62*(5), 728–754.

Khan, H. Z., Bose, S., Mollik, A. T., & Harun, H. (2020). "Green washing" or "authentic effort"? An empirical investigation of the quality of sustainability reporting by banks. *Accounting, Auditing & Accountability Journal*, 1–39.

Lane, E. L. (2013). Greenwashing 2.0. *Columbia Journal of Environmental Law*, *38*, 279.

Laufer, W. S. (2003). Social accountability and corporate greenwashing. *Journal of Business Ethics*, *43*(3), 253–261.

Lee, M. T., & Suh, I. (2022). Understanding the effects of environment, social, and governance conduct on financial performance: Arguments for a process and integrated modelling approach. *Sustainable Technology and Entrepreneurship*, *1*(1), 100004.

Li, T. T., Wang, K., Sueyoshi, T., & Wang, D. D. (2021). ESG: Research progress and future prospects. *Sustainability, 13*(21), 11663.

Loew, E., Klein, D., & Pavicevac, A. (2019). *Corporate social responsibility reports of European banks—An empirical analysis of the disclosure quality and its determinants* (European Banking Institute Working Paper Series No. 56).

Lyon, T. P., & Montgomery, A. W. (2015). The means and end of greenwash. *Organization & Environment, 28*(2), 223–249.

Macaire, C., & Naef, A. (2021). Greening monetary policy: Evidence from the People's Bank of China. *Climate Policy*, 1–12.

Marciniak, A. (2009). Greenwashing as an example of ecological marketing misleading practices. *Comparative Economic Research. Central and Eastern Europe, 12*(1/2), 49–59.

Marquis, C., Toffel, M. W., & Bird, Y. (2015). *Scrutiny, norms, and selective disclosure: A global study of greenwashing* (Forthcoming in Organization Science, Harvard Business School Organizational Behavior Unit Working Paper, 11–115).

Marquis, C., Toffel, M. W., & Zhou, Y. (2016). Scrutiny, norms, and selective disclosure: A global study of greenwashing. *Organization Science, 27*(2), 483–504.

Migliorelli, M., & Dessertine, P. (2019). *The rise of green finance in Europe: Opportunities and challenges for issuers, investors and marketplaces*. Palgrave Macmillan.

Mitchell, L., & Ramey, W. (2011). Look how green I am! An individual-level explanation for greenwashing. *Journal of Applied Business and Economics, 12*(6), 40–45.

Nedopil Wang, C., Lund Larsen, M., & Wang, Y. (2020). Addressing the missing linkage in sustainable finance: the 'SDG Finance Taxonomy'. *Journal of Sustainable Finance & Investment*, 1–8.

Pacces, A. M. (2021). Will the EU taxonomy regulation foster sustainable corporate governance? *Sustainability, 13*(21), 12316.

Parguel, B., Benoît-Moreau, F., & Larceneux, F. (2011). How sustainability ratings might deter 'greenwashing': A closer look at ethical corporate communication. *Journal of Business Ethics, 102*(1), 15–28.

Parguel, B., Benoit-Moreau, F., & Russell, C. A. (2015). Can evoking nature in advertising mislead consumers? The power of 'executional greenwashing'. *International Journal of Advertising, 34*(1), 107–134.

Pérez, A. (2015). Corporate reputation and CSR reporting to stakeholders: Gaps in the literature and future lines of research. *Corporate Communications: An International Journal, 20*(1), 11–29.

People's Bank of China (PBOC). (2021). *Green Bond Endorsed Projects Catalogue (2021 Edition)*. http://www.pbc.gov.cn/goutongjiaoliu/113456/113469/4342400/2021091617180089879.pdf

Phillips, M. (2019). "Daring to care": Challenging corporate environmentalism. *Journal of Business Ethics, 156*(4), 1151–1164.

Seele, P., & Gatti, L. (2017). Greenwashing revisited: In search of a typology and accusation-based definition incorporating legitimacy strategies. *Business Strategy and the Environment, 26*(2), 239–252.

Shoaf, V., Jermakowicz, E. K., & Epstein, B. J. (2018). Toward sustainability and integrated reporting. *Review of Business, 38*(1), 1–15.

Siano, A., Vollero, A., Conte, F., & Amabile, S. (2017). "More than words": Expanding the taxonomy of greenwashing after the Volkswagen scandal. *Journal of Business Research, 71*, 27–37.

Terrachoice. (2007). *The "Six Sins of GreenwashingTM": A study of environmental claims in North American consumer markets.* https://sustainability.usask.ca/documents/Six_Sins_of_Greenwashing_nov2007.pdf

Tripathy, A., Mok, L., & House, K. (2020). Defining climate-aligned investment: An analysis of sustainable finance taxonomy development. *The Journal of Environmental Investing, 10*(1).

Troeger, T. H., & Steuer, S. (2021). *The role of disclosure in green finance* (European Corporate Governance Institute-Law Working Paper, 604).

Vaze, P., Meng, A., & Giuliani, D. (2019). *Greening the financial system.* Climate Bonds Initiative.

Vollero, A., Palazzo, M., Siano, A., & Elving, W. J. (2016). Avoiding the greenwashing trap: Between CSR communication and stakeholder engagement. *International Journal of Innovation and Sustainable Development, 10*(2), 120–140.

Vos, J. (2009). Actions speak louder than words: Greenwashing in corporate America. *Notre Dame Journal of Law, Ethics & Public Policy, 23*, 673.

Watson, B. (2017). The troubling evolution of corporate greenwashing. *Chain Reaction, 129*, 38–40.

Wu, Y., Zhang, K., & Xie, J. (2020). Bad greenwashing, good greenwashing: Corporate social responsibility and information transparency. *Management Science, 66*(7), 3095–3112.

Yang, Z., Nguyen, T. T. H., Nguyen, H. N., Nguyen, T. T. N., & Cao, T. T. (2020). Greenwashing behaviours: Causes, taxonomy and consequences based on a systematic literature review. *Journal of Business Economics and Management, 21*(5), 1486–1507.

Yu, E. P. Y., Van Luu, B., & Chen, C. H. (2020). Greenwashing in environmental, social and governance disclosures. *Research in International Business and Finance, 52*, 101192.

Zhang, D. (2022). Environmental regulation and firm product quality improvement: How does the greenwashing response? *International Review of Financial Analysis, 80*, 102058.

CHAPTER 6

Financing the Green Recovery: The New Directions of Finance After the COVID-19 Crisis

6.1 Introduction

The economic fallout of COVID-19 has severely impacted societies and economies around the world. The shock waves were caused not only by the increased fatalities worldwide, but also by the disease control measures adopted in the infected nations, which entailed considerable disruption and contraction in economic activity. This resulted in sudden, sharp declines in government and business revenues, and loss of jobs. Apart from the enormous impact on healthcare systems, economic and social life were also affected by the pandemic (Clemente-Suárez et al., 2021). Lockdowns led to a significant decline in the GDP of infected nations, with devasting consequences expected in the following years (Altig et al., 2020; Jena et al., 2021). Indeed, the pandemic raised concerns about the dangers of decades of prosperity that failed to take responsibility for future generations; it has exacerbated those vulnerabilities, created new ones, and reduced people's ability to cope with shocks (Vitenu-Sackey & Barfi, 2021).

The policy responses to the pandemic have been strictly related to the causes of the crisis; thus, overcoming the negative effects required governments to prepare long-term recovery and stimulus packages to support economic growth and employment security. The linkage is becoming more evident between health impacts, natural disasters, pandemics such as COVID-19, and changes in the climate, oceans, and forests (Akrofi et al.,

© The Author(s), under exclusive license to Springer Nature
Switzerland AG 2022
A. Rizzello, *Green Investing*, Palgrave Studies in Impact Finance,
https://doi.org/10.1007/978-3-031-08031-9_6

2021; Mishra et al., 2021). For these reasons, the common trend seen in the massive economic policy response to the COVID-19 pandemic is the greening of economies, as well as of the financial system. The European Union area is leading the policy innovation toward greenness and sustainability by turning this immense challenge into an opportunity, with the provision of recovery funds oriented to a *green new deal* (Pettifor, 2020). Indeed, the "green driver" in the various forms of national recovery plans has been suggested not only by academics (for an overview, see Barbier, 2020; Lahcen et al., 2020; Le Billon et al., 2021; Mukanjari & Sterner, 2020; Werikhe, 2022), but also by international policy players, such as OECD (Agrawala et al., 2020) or the United Nations (O'Callaghan and Murdock, 2021).

Within this context, recovery strategies focused also on transition activities (Cazcarro et al., 2022; Pianta et al., 2021), which led to supporting a growth in fossil fuel or carbon-intensive investments that diverged from the Paris Agreement's trajectory. Conversely, there is a critical window of opportunity to reset and re-think the concept of sustainable development (Leach et al., 2021), as well as the financing of such activities. The co-occurrence of massive policies to stimulate green investments, market regulatory transformations, and green financial innovations, form a basis for accelerating *green-led* paradigm shifts in financial theories and practices, by providing a unique opportunity for scholars and practitioners to contribute to designing a sustainable financial system.

The remainder of this chapter is organized as follows. Section 6.2 explores the nexus between the pandemic and the need for green stimulus packages. Section 6.3 presents an overview of the main characteristics of the post-COVID recovery actions and their potential to boost a sustainable financial market. Section 6.4 explores the emerging conceptual dimensions that are accelerating a re-think of the theoretical concepts of finance and green sustainability. Section 6.5 highlights the practical implications of this transformative trend, and Sect. 6.6 suggests future avenues of research to investigate the new directions of finance imposed from the magnitude of such green response. Finally, Sect. 6.7 concludes.

6.2 The Nexus Between the Pandemic Crisis and Green Recovery Plans

Long-dominant development models, which promoted social and economic change through unlimited growth, carbon-intensive industries, and market liberalization, were undoubtedly threatened by the COVID-19 spread (Callegari & Feder, 2022). Reflections from many perspectives regarding the causes and effects of the pandemic crisis influenced the response actions. Indeed, this unprecedented challenge exposed the high vulnerability of mainstream approaches to the development concept and its economic archetypes, with diverse regions and social groups calling for solutions to radically transform future post-pandemic economies (Kołodko, 2020; Laranja & Pinto, 2022; Moreno et al., 2021; Song & Zhou, 2020).

On the other hand, stimulus reaction strategies worldwide drew attention to incentives for further development of the green and circular economy, as well to promoting additional investment in renewable energy projects. In brief, a wide range of diverse voices (for instance, Corfee-Morlot et al., 2021; D'Orazio, 2021; Fears et al., 2020; Pinner et al., 2020) quickly began to call on government policies to think beyond a pure economic recovery, in order to rescue jobs and livelihoods, and to reinvigorate economic growth. Thus, many countries heeded the call by committing their repairing efforts to a "green recovery" through national or supranational stimulus packages (Andrijevic et al., 2020; Bongardt & Torres, 2022; Boston, 2020; Gusheva & de Gooyert, 2021). However, the nexus between the pandemic crisis and such responses needs to be clarified.

The roots of green COVID-19 responses may be identified through an analysis of the pandemic crisis and its impacts. For instance, the structural drivers that led to the disease's emergence and spillover have been identified as an intersecting, structural interaction of political, social, and ecological processes (Farzanegan et al., 2021; Mogi & Spijker, 2021; Perveen et al., 2021). In more detail, the COVID-19 spread has been considered as the last (but the largest) disease outbreak provoked by a non-linear human-animal-ecological dynamic, produced by a structural, political, and economic development approach that had little respect or consideration for dynamic ecological consequences (Fuentes et al., 2020). Thus, with the rise in both old and emergent inequalities and vulnerabilities, COVID-19 focused attention on the fragility of those economies that fail to consider the interactions and intersections of social, economic, political, and environmental dimensions (Kalfagianni & Papyrakis, 2022; Kelly, 2021; Kumar & Ayedee, 2021; Newell & Dale, 2020; Zabaniotou, 2020).

Consequently, the need to reflect on post-pandemic transformations produced a triple, and partially overlapping and intersecting, effect of re-thinking the structural archetypes of: (i) economic growth and development, (ii) sustainability, and (iii) the economy and financial system—each of which centered on environmental protection and the provision of additional environmental benefits (Henry et al., 2020; Jamaludin et al., 2020; Taherzadeh, 2021; Tian et al., 2022). Indeed, if the concepts of economic growth should consider human–nature interactions in order to prevent pandemic crises, then the protection of future generations' health should be integrated as one of the three sustainability pillars. Indeed, environmental protection implies the protection of future generations' health, as a way of enhancing society's risk-anticipation capacities (Guerriero et al., 2020; Ranjbari et al., 2021).

Regarding the impact on the rebuilding of economic models and financial system archetypes, the pandemic highlighted different forms of actions that incorporate and integrate environmental risk factors within the mainstream models. Indeed, many central banks developed measures to include pandemics as one of the top risks for economic systems via an environmental crisis, given that environmental change may infect financial stability through environmental change-related events, such as pandemics (Bahaj & Reis, 2020; Gortsos, 2020; Stiglitz et al., 2020). On the other hand, the focus on mitigating environmental risks within the financial sector also produced efforts to prevent and balance the effect of transition-related risks from the accelerated shift toward low-carbon activities. For instance, this acceleration could produce crises deriving, for instance, from the systemic consequences for financial markets produced by the crisis in the fossil fuel-based industry (for instance, Kedward et al., 2020; Rudebusch, 2021).

For these reasons, suggestions for the green orientation of the massive stimulus packages, in order to stimulate the post-pandemic economies, also drew on considerations about the recovery attempts' limited ability to re-capitalize the global banking system after the 2008 financial crisis; they focused on fixing existing structures characterized by a low degree of *anti-fragility* (Taleb & Douady, 2011). Therefore, in economic responses to COVID-19, the majority of suggestions concern the possibility of enhancing resilience, and promoting the green transition of economies and financial systems (Bryce et al., 2020; Steffen et al., 2020; Zenghelis, 2021). Briefly, the nexus of the pandemic and policy discussion on the need for green stimulus packages has been influenced by the debate on re-thinking economic growth and the sustainability of economies and financial systems. This is encouraging wider debates on: (i) re-thinking developments, (ii) resilience of economic and financial systems, and (iii) finance, as tools to accelerate the green transition.

6.3 An Overview of the Main Green Post-COVID Recovery Actions

By April 2021, approximately USD 16 trillion had been reported globally as the public resources invested in responding to COVID-19's negative impacts; most of this amount was rescue spending and health funding to overcome the worst of the crisis (International Monetary Fund, 2021). After the first emergency phase, more countries progressed from rescue spending toward announcing their recovery packages. For instance, the COVID-19 expenditure responses can be grouped into crisis management interventions (such as lockdowns and restrictions, and related recovery funds), pandemic preparedness (e.g., pandemic management protocols), and stimulus measures, mainly economic and financial (Alberola et al., 2021). However, as of the end of January 2022, the Global Recovery Observatory of Oxford University found that the amount of spending on green measures represented only 31.2% of total COVID-19 recovery spending (USD 3,847 billion) announced worldwide (O'Callaghan et al., 2022). A significant amount of green measures appear in the budget of the EU Recovery and Resilience Facility allocated to EU Member States, whose Recovery and Resilience Plans (or RRPs) are required to allocate at least 37% of funding to climate action goals (Vanhercke & Verdun, 2022). On the other hand, according to OECD/IEA (2021), government support measures for fossil fuels amounted to USD 345 billion in the year 2020. More recently, the US government also announced the American Jobs Plan, which could substantially shift the dial in the green direction.

By examining the types of actions, it can be seen that grants/loans represent the favorite measure introduced by countries, accounting for around 39% of the total measures with clear environmental implications registered in the OECD Green Recovery Database,[1] followed by tax reductions and other subsidies. According to this database, the energy, green buildings, and ground transport sectors are the largest beneficiaries of green recovery measures, with a dedicated budget, while most of the remaining resources are dedicated to *economy-wide* policies. In terms of environmental targets included in recovery measures, 90% address climate change mitigation, while air pollution accounts for 64% of environmental targets addressed by green COVID-19 measures.

In addition to green measures related to investment recovery, a large number of identified regulatory measures have important implications for economic trajectories, and the majority have been recognized as benefiting the environment (Cazcarro et al., 2020). As a result, the policy efforts to inject a green-led investment and regulatory stimulus still present a mixed picture, given that according to the aforementioned OECD Database, two-thirds of the remaining investments are directed to activities with neutral or negative environmental impacts. In this sense, the target expressed with the words "build back better" (Furman, 2021) requires further urgent decisions to be made, to incorporate a longer-term perspective in the COVID-19 responses.

A Focus on the European Union Area

In December 2020, the EU announced its agreement on the EU budget and Next Generation EU (NGEU) recovery package (Hinarejos, 2020). This set of fiscal remedies aims to provide additional spending of EUR 750 billion in total, financed by borrowing at the EU level (Brunsden et al., 2020). Through a special Recovery and Resilience Facility (RRF), the funds are channeled to the Member States in the form of grants (EUR 390 billion) and loans (EUR 360 billion), to be committed between 2021 and 2023 (Echebarria Fernández, 2021). Overall, 30% of the NGEU budget will be targeted at climate change-related spending (Sikora, 2021). The RRF entered into force in February 2021, and at the end of June 2021, EU member states submitted their Recovery and Resilience Plans (RRP) to the European Commission (Verdun et al., 2022). Notably, the European Commission President Ursula von der Leyen announced that 30% of the EUR 750 billion for the Next Generation EU budget

---

[1] Available at https://www.oecd.org/coronavirus/en/themes/green-recovery.

will be raised through green bonds (European Commission, 2020). It was also indicated that 37% of the funding will be invested in European Green Deal objectives, the set of proposals that align the EU's climate, energy, transport, and taxation policies with reducing net greenhouse gas emissions by at least 55% by 2030, compared to 1990 levels. In the same year, the European Commission's Technical Expert Group (TEG) on Sustainable Finance (EU TEG, 2020) published five high-level principles for recovery and resilience, and detailed information on implementing the EU Taxonomy in EU recovery planning. These can be summarized as follows:

1. Measures focusing on the principle "building back better," to meet at least 50% of the emission reductions needed by 2030;
2. Building resilience into everything in order to increase preparedness for future shocks, by encouraging investments in: (i) social resilience (actions to protect people at risk), (ii) economic resilience (by privileging sectors that contribute most to a more sustainable and resilient economy), and (iii) ecosystem resilience, directed to protecting and rebuilding natural capital and biodiversity in Europe;
3. Ensuring that the recovery plans are in line with the "do no harm" principle;
4. Adding private finance to public funds, by creating stable and robust frameworks to attract private investment;
5. Incentivizing international partnerships to avoid the failure of single states caught in a shock.

Finally, in October 2020, the European Commission announced the issuing of the first EU SURE bond of up to EUR 100 billion, aligned with Social Bond Principles (SBP), to help Member States cover the costs directly related to financing national measures in response to the pandemic (Crowe, 2021).

6.4 Theoretical Innovations in Financial Studies

Moving from different theoretical perspectives, also in Academia a group of "alternative voices" recognized interconnections between the environment, development, and finance. They all converged toward a set of pre-paradigmatic assumptions for a sustainable/green-led paradigm shift in finance (for an overview, see Climate Risk Is Investment Risk, 2020; Faber, 2018; Lagoarde-Segot et al., 2021; Langley, 2020; Luisetti, 2019; Marques, 2020; Murshed, 2020; Ryszawska, 2018; Schwab, 2019; Stensrud & Eriksen, 2019; Taleb, 2012; Yanovski et al., 2020). These academic tendencies can be clustered into three main types:

- The affirmation of an environmentally driven proposition in financial theory;
- The progressively shifting focus from resilience to the anti-fragility function of finance;
- The decline of short-termism.

Toward an Environmentally Driven Proposition in Financial Theory

The increasing climate of uncertainty, deriving from repeated economic and financial shocks, started in the twenty-first century with the Twin Towers attack, followed by the 2008 financial crisis, up to the recent wave produced by the pandemic. This has brought a growing consensus that the socio-economic system is not a separate entity from the financial system (for instance, de Vries, 2019; Monasterolo, 2020; Shiller, 2013; Storm, 2018), and the consequent need to move beyond neoclassical financial theory. Based on these considerations, the need for radical change in the financial system to face these challenges was recognized by not only a large number of multilateral institutions, but also a consistent body of research (Lagoarde-Segot, 2019; Lagoarde-Segot & Paranque, 2018; Loorbach et al., 2020; Newell, 2022; Paranque & Pérez, 2016; Penna et al., 2021; Perez, 2021). These studies revealed the incompatibility of theoretical neoclassical financial assumptions with the occurrence of this unexpected social, ecological, and financial crisis.

In particular, by unveiling the weaknesses of the neoclassical financial model adopted over the past 50 years, this group of studies converged toward a new vision that considers the financial system as part of a broader interconnected system, with complex interactions between the financial, socio-economic, and the biological sphere (Lagoarde-Segot & Martínez, 2021). Beyond recognizing the validity of this new theoretical proposition, these growing alternative voices drew attention to the necessity for innovation in financial theory, by acknowledging the role of values, ethics, and ideologies in finance studies (Maida, 2021; San-Jose et al., 2021; Schäfer, 2019). The consideration and integration of the environmental elements within financial theory propositions allows the scientific community to incorporate into their analyses and prescriptions the complex interactions between financial systems, social provisioning, and ecological systems.

From Resilient to Anti-fragility Finance

The COVID-19 pandemic has highlighted an urgent need to consider resilience in finance (OECD, 2020), both in the financial system itself (Hynes et al., 2020) and in the role played by capital and investors in making economic and social systems more dynamic and able to withstand external shocks (Cumming et al., 2021). Indeed, the Coronavirus crisis has identified vulnerabilities of economic and financial systems that have arisen, in part due to unintended consequences, by highlighting the system's insufficient capacity to ensure sustainable and inclusive economic growth (Stevano et al., 2021; Wullweber, 2020).

However, the goal of finding an enabling function to incentivize resilience is not new in the green finance sector (see, for instance, the manifesto of the United Nations Environmental Programme, 2011), nor in financial studies (e.g., Boissinot et al., 2016; Brassett & Holmes, 2016; Meltzer, 2016). This represents a further stream of research derived from the academic debate on the concept of socio-ecological resilience, in relation to uneven capitalist development, conflicts, and ecological crises (see, among others, Holling, 1986). Interestingly, however, the recurring theme of socio-ecological *crisis* in contemporary development discourse also aligns with a renewed interest in resilience thinking for finance among scholars and policymakers; this can be contextualized within the debate on how to prevent new economic and financial shocks similar to those of the COVID-19 pandemic (Hynes et al., 2020; Johnston et al., 2021). However, by interconnecting the policy (OECD, 2020; World Bank Group, 2019) and academic debates (Goodell, 2020; Wei & Han, 2021) about the transformation of financial system by integrating mechanisms for absorbing and neutralizing shocks derived from socio-ecological crisis, one can gain a more nuanced view of the role of finance as a tool to overcome such shocks. In simplified terms, while the resilience concept identifies the system's ability to absorb changes and still survive, the need emerging from the COVID-19 crisis calls for systems to go beyond resilience or stability; this is well summarized in the concept of "anti-fragile" (Taleb, 2012). Indeed, an anti-fragile system performs better than a resilient one which resists shocks and stays the same, thanks to the former's capacity to benefit from harm by *learning*—as in the case of the immunization system, in line with the Shumpetarian's view (1943) of the positive link between disorders and the benefit gained from disorders.

In this light, developing anti-fragility capacity represents the perfect preparation for surprising and unexpected future events (*Black Swan*) that may threaten economic systems. In a broader sense, the resilience-led roadmap may not ensure sustainability without an institutional framework that enables economic actors to deal with the uncertainties and threats. Thus, the focus on developing an anti-fragility *immunization* that helps the economic and financial system to benefit from uncertainty, randomness, and disorder, seems more appropriate than the discourse around resilience, which is focused, instead, on stability and conservation of the economic and financial system. For these reasons, the recognition of the anti-fragility function of finance will be able to achieve the aim of perfecting its role for incentivizing sustainability outcomes, in terms of enabling systems not only to survive, but also to prevent collapse. For these reasons, introducing an institutional framework for developing indicators of fragility may enable the financial system to identify future Black Swan events, and prepare for them.

The Progressive Decline of Short-Termism

According to Marginson and McAulay (2008), short-termism refers to corporate and financial markets' prioritization of near-term shareholder interests over long-term growth; it has been classified as one of the main barriers to the sustainability of a corporate strategy. Similarly, within the financial sector, short-termism relates to the failure to adequately account for and invest in long-term environmental, social, and economic sustainability.

The topic gained attention globally after the 2008 financial crisis, which evidenced the interconnection between short-termism and systemic risk (Burkart & Dasgupta, 2021; Sanderson et al., 2019; Yanovski et al., 2020). The degenerative practice of short-termism covers multiple financial activities such as trading and leverage buyout practices, as well as investing based on a willingness to renounce the long-term value creation of their investment. The investors' preference for the immediate high returns without accounting for long-term implications may lead to underperformance for long-term investors, and, by extension, underperformance of the economy as a whole. For these reasons, the short-termism debate has prompted the call for a paradigmatic shift, among financial studies (Davies et al., 2014; Fried & Wang, 2019; Nesbitt, 2009; Verstein, 2017; Walsh & O'Riordan, 2020). In other words, the destruction of long-term value through a short-term approach to finance compromises the achievement of sustainability outcomes and incentivizes the recurrence of unethical practices, which were the basis of the recent financial crisis. Furthermore, the destruction of long-term value represents an unpaid cost of a transaction, defined in financial theory by the expression "(negative) externalities" (Mählmann, 2022; Ziolo et al., 2019), which is often considered a sign of market failure (Gillingham & Sweeney, 2010). In the case of investment, such externalities are temporally "distant" from the time of the transaction, and for this reason they are not considered in investment decision-making (Bonnefon et al., 2019; Marlow et al., 2011). Furthermore, as a paradoxical implication, the voluntary inclusion of future externalities in business practice produces a market penalty for those actors who increase their cost of transactions. Environmental damages, such as climate change, are a perfect example of this (Libecap, 2014).

In this regard, the conceptualization of investment strategies that aim to incorporate long-term values into investment decisions represents a form of financial innovation (Shiller, 2004) that is best adapted to prevent the negative externalities of an investment. Also recognized with the term "patient capital" (Deeg & Hardie, 2016), this approach has the potential to guide the financial system back in the direction of long-term investing by encouraging less pro-cyclical investment strategies. Within this arena, impact investing (II) is the investment strategy intended to simultaneously obtain financial as well as environmental/social measurable returns from an investment (for an overview, see Barber et al., 2021; Brest & Born, 2013; Chiappini, 2017). Indeed, this may be the keystone for integrating long-term value creation with investment decisions. This is particularly true for the green finance sector, where the mitigation of negative externalities for the environment is mainly achieved through taxonomies and portfolios' green-alignment techniques. Thus, environmental impact investing practice should lead financial actors to consider green finance not as an objective in itself, but rather as a tool to improve environmental conditions. By adding this element into a green investment strategy, the focus is on the potential *impact* of green investments, rather than on *what* green investment is directed to fund. The extension and diffusion of this transformative strategy in the green finance sector could produce relevant theoretical as well as practical implications that can be usefully explored, understood, and disseminated.

6.5 Implications for the Greening of the Financial Sector

From the emergence of the conceptual dimensions highlighted above, we can derive practical implications of transforming the green finance market after its accelerated evolution, assisted by responses to the negative impact of the pandemic. In particular, the affirmation of an environmentally driven proposition in financial theory, the progressively shifting focus from resilience to the anti-fragility function of finance, and the decline of short-termism, imply practical consequences of this transformative phase of the financial sector. In other words, an integrated social-ecological system perspective is needed to address the discrepancy between the emerging practices in green investments and business at the micro level, and the outcomes at the macro level. These practical consequences can be grouped, respectively, in terms of:

- The complete inclusion, within the systemic risks, of climate and environmental risks;
- The inclusion of finance for *transition* within financial green-targeting investments;
- The definitive integration of environmental factors into modern frameworks of investors' fiduciary duties.

Integration of Climate and Environmental Risks for Financial Stability

The inclusion of an environmental proposition within financial value creation implies, among other outcomes, the systematic consideration of risk induced by climate change and environmental disequilibrium. In contrast to the traditional forms of risk within the financial sector, such risk is considered systemic (Sutton, 2019) by reverberating its effects across regions and sectors, through interconnected socioeconomic and financial systems (Undorf et al., 2020). In other words, such risks arise from the interaction between the natural environment and complex human systems; this raises a danger that the entire system may break down (De Amorim et al., 2018), given the high level of interdependence and complexity in the social systems (Reyers et al., 2018) of the so-called Anthropocene era (Hughes et al., 2017). For these reasons, once they have emerged, such risks pose a serious threat to the structure and function of human society (Biesbroek et al., 2014); but at the same time, policy changes to address climate-related threats produce market reactions (Birindelli & Chiappini, 2021). In this light, the existing literature highlights the importance of systemic risk induced by climate change (among others, Battiston et al., 2021; Monasterolo, 2020); however, there are still deficiencies in channeling its dynamics and assessing the risk within financial institutions.

Based on these considerations, financial institutions need to accurately assess the climate and environmental risks to which they are exposed, given that these risks lead to the excessive allocation of financial resources to polluting or high-carbon sectors. Beyond exposing financial institutions to non-compliance risks (if they include a maximum percentage of alignment to carbon sectors), the underestimation of such risks threatens financial institutions' own balance sheets and financial stability. Such effects could be produced, for instance, by affecting the resilience of their business model over the medium to longer term—especially for those heavily engaged in sectors and markets which are particularly vulnerable to climate-related and environmental risks (Chenet et al., 2021; Ranger et al., 2021). Furthermore, the risks can simultaneously represent drivers of several different risk categories, and sub-categories of existing ones (Braun et al., 2020). Thus, the importance of financial institutions considering and assessing climate-related and environmental risk when formulating and implementing their business strategy and risk management frameworks will become a primary focus, especially after the COVID-19 crisis is finally contained and a transition to a green economy is imperative.

Environmental Alignment and Transition Finance

The progressively shifting focus from resilience to the anti-fragility function of economic and financial systems is affirmed by learning lessons from the crisis. The negative impact of the pandemic influenced this change, and, as a consequence of response policies, unprecedented recovery stimuli were directed to financing facilities in different sectors of the global economy, with a significant amount proactively funding the transition to fulfill the Paris Agreement on climate change and the UN Sustainable Development Goals (SDGs).

Defined as the journey necessary to achieve sustainable outcomes (Piemonte et al., 2019), the concept of transition is not new within economic discourses, although there are different interpretations of the aim of achieving an improved level of income. However, more recently the concept has been adopted to define the green transition (OECD, 2021), as the road to transforming economies with differing endowments and structures to meet global net-zero emissions.

Within this context, there is a debate on the definition of a just transition (Fernandes et al., 2021). A particular challenge arises if transition refers not only to scaling-up zero or near-zero emitting technologies and businesses, but also to creating feasible emission reduction pathways. Indeed, highly CO_2-emitting economic activities are likely to continue in certain jurisdictions, while declining in others. There is therefore an urgent need to withhold financial support for high-emitting and intransigent sectors, in order to achieve a global net-zero CO_2 emissions target (Khurshid & Deng, 2021).

In this context, the concept of *transition finance* emerged, intended to best capture the holistic approach to financing green growth (Piemonte et al., 2019); however, it includes various conceptual shades and its definition extends beyond the *use of proceeds* in green finance (Caldecott, 2020). For these reasons, a holistic approach to financing green growth should not be reduced only to *finance for green virtue*, as this would dangerously diverge from a systemic change by completely isolating carbon industries in their transformation journey.

To address this grey area associated with transition finance, regulatory measures should allow green capitals to flow to companies that have the potential to meaningfully contribute to climate stabilization and the transition to a zero-carbon future, by transforming their inherently carbon-intensive business models. For instance, recognizing these activities in taxonomies would require additional and more inclusive transitional categories, to better represent the different transitions underway by excluding activities that significantly harm the taxonomy's environmental objectives. The debate around measures enabling transition finance is at an initial stage, but it is notable that continuing debates about integrity (for instance, strict standards) could confine green finance to a niche position in the global finance industry, and undercut its important role in capital allocation.

Including Environmental Factors into Fiduciary Duties Frameworks

The concept of fiduciary duty is not new in finance; it relates to asset managers' role in managing money on behalf of beneficiaries and savers, based on trust, confidence, and ethics (Lydenberg, 2014). The concept of fiduciary duty relates to the decision-making process rather than the obligation to achieve a particular outcome. In the context of green finance, the concept relates essentially to the inclusion of ESG opportunities and risks in investment practices and processes. The aim of long-term value creation in green investing requires the incorporation of these commitments into investment mandates, and reporting the results of their implementation (Gary, 2019). However, these values are often expressed as financial numbers and ratios, such as net present value; therefore, such reasoning transforms the non-financial into the financial. Indeed, the conceptual debate on including ESG issues as a requirement of investors' duties and obligations is now over, but further work is required to establish environmental impact assessment within the financial world.

In other words, the debate should move beyond the evaluation of companies' environmental performance via ESG research, toward assessing the environmental impact of the green-aligned investment. For this purpose, the practice derived from impact investing may be useful for introducing such forms of return into green finance investments. However, the impact investing field has been concerned with the assessment and evaluation of social impact rather than environmental impact, and so far, relatively little research has investigated the impact of green investments. In more detail, the progress in measuring environmental impact is limited, but there is a consolidated set of tools to identify indicators and targets. On the other hand, mechanisms enabling the creation of environmental impact still need to be explored and understood.

6.6 Future Lines of Research

The debate around the transformation of finance after COVID-19 will involve scholars in the investigation of innovative perspectives addressing different approaches. On a macro level, the intersection of economic and financial system with the ecological sphere, will require cross-sectoral studies addressing the topic of *ecological finance* within an *integral ecology* perspective through grounded theory or conceptual framework design. On a meso level, the exploration of anti-fragility financial mechanisms, instruments, and approaches should be identified with quantitative as well as qualitative approaches in different contexts and markets, with a particular attention to the European Union area that appears more advanced in green financial policies and regulation. On a micro level, future studies should provide empirical evidence about the management of environmental risks within financial institutions and asset managers as well as the exploration of the overcoming of short-termism in the field of behavioral finance and portfolio theories.

6.7 Conclusion

The chapter addresses the issue of the evolution of finance after COVID-19 pandemic. Through a triangulated analysis derives a framework that sees the interaction of three different transformative factors, as synthetically illustrated in Fig. 6.1, and outlines future lines of research.

The negative impact produced by the pandemic and related stimulus responses met the contemporary evolution in classical economic and financial assumptions. More simply, the green orientation impressed in

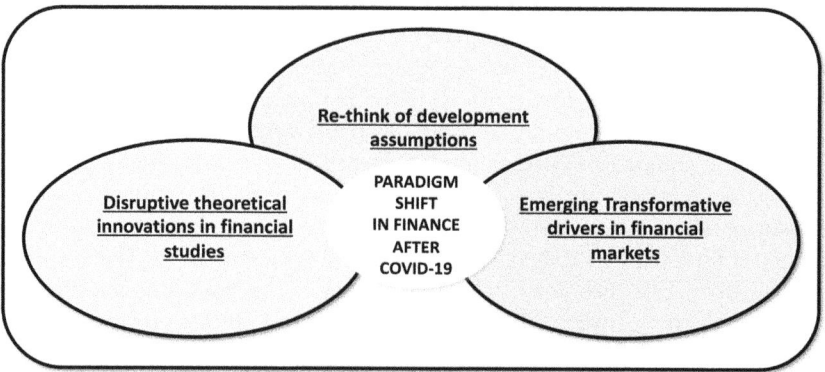

Fig. 6.1 Determinants for the new role of finance after COVID-19 (*Source* Author's elaboration)

such packages intersects determined innovations and re-thinking of classical assumptions in economic and financial studies. The marriage of such transformation converges toward a disruptive nuance for financial sector in the next years. Academic research will serve as a determinant actor in the decoding of this *new normal* for financial sector.

References

Agrawala, S., Dussaux, D., & Monti, N. (2020). *What policies for greening the crisis response and economic recovery?: Lessons learned from past green stimulus measures and implications for the COVID-19 crisis.*

Alberola, E., Arslan, Y., Cheng, G., & Moessner, R. (2021). Fiscal response to the COVID-19 crisis in advanced and emerging market economies. *Pacific Economic Review, 26*(4), 459–468.

Akrofi, M. M., Mahama, M., & Nevo, C. M. (2021). Nexus between the gendered socio-economic impacts of COVID-19 and climate change: Implications for pandemic recovery. *SN Social Sciences, 1*(8), 1–20.

Altig, D., Baker, S., Barrero, J. M., Bloom, N., Bunn, P., Chen, S., Davis, S. J., Leather, J., Meyer, B., Mihaylov, E., Mizen, P., Parker, N., Renault, T., Smietanka, P., & Thwaites, G. (2020). Economic uncertainty before and during the COVID-19 pandemic. *Journal of Public Economics, 191*, 104274.

Andrijevic, M., Schleussner, C. F., Gidden, M. J., McCollum, D. L., & Rogelj, J. (2020). COVID-19 recovery funds dwarf clean energy investment needs. *Science, 370*(6514), 298–300.

Bahaj, S., & Reis, R. (2020). Central bank swap lines during the Covid-19 pandemic. *Covid Economics, 2*(8).

Barber, B. M., Morse, A., & Yasuda, A. (2021). Impact investing. *Journal of Financial Economics, 139*(1), 162–185.

Barbier, E. B. (2020). Greening the post-pandemic recovery in the G20. *Environmental and Resource Economics, 76*(4), 685–703.

Battiston, S., Dafermos, Y., & Monasterolo, I. (2021). Climate risks and financial stability. *Journal of Financial Stability, 54*, 100867.

Biesbroek, G. R., Termeer, C. J., Klostermann, J. E., & Kabat, P. (2014). Rethinking barriers to adaptation: Mechanism-based explanation of impasses in the governance of an innovative adaptation measure. *Global Environmental Change, 26*, 108–118.

Birindelli, G., & Chiappini, H. (2021). Climate change policies: Good news or bad news for firms in the European Union? *Corporate Social Responsibility and Environmental Management, 28*(2), 831–848.

Boissinot, J., Huber, D., & Lame, G. (2016). Finance and climate: The transition to a low-carbon and climate-resilient economy from a financial sector perspective. *OECD Journal: Financial Market Trends, 2015*(1), 7–23.

Bongardt, A., & Torres, F. (2022). The European green deal: More than an exit strategy to the pandemic crisis, a building block of a sustainable european economic model. *JCMS: Journal of Common Market Studies, 60*(1), 170–185.

Bonnefon, J. F., Landier, A., Sastry, P., & Thesmar, D. (2019). *Do investors care about corporate externalities? Experimental evidence* (HEC Paris Research Paper No. FIN-2019-1350).

Boston, J. (2020). Transforming the economy: Why a 'green recovery' from Covid-19 is vital. *Policy Quarterly, 16*(3).

Brassett, J., & Holmes, C. (2016). Building resilient finance? Uncertainty, complexity, and resistance. *The British Journal of Politics and International Relations, 18*(2), 370–388.

Braun, B., Claeys, I., de Arriba-Sellier, N., Fontan, C., Krahé, M., Pistor, K., Schreur, V., Smoleńska, A., Solana, J., & van't Klooster, J. (2020). *General Comments from the Expert Group on Draft ECB Guide on climate-related and environmental risks*.

Brest, P., & Born, K. (2013). When can impact investing create real impact. *Stanford Social Innovation Review, 11*(4), 22–31.

Brunsden, J., Fleming, S., & Khan, M. (2020). EU recovery fund: How the plan will work. *Financial Times*, 21.

Bryce, C., Ring, P., Ashby, S., & Wardman, J. K. (2020). Resilience in the face of uncertainty: Early lessons from the COVID-19 pandemic. *Journal of Risk Research, 23*(7–8), 880–887.

Burkart, M., & Dasgupta, A. (2021). Competition for flow and short-termism in activism. *The Review of Corporate Finance Studies, 10*(1), 44–81.

Caldecott, B. (2020). Defining transition finance and embedding it in the post-Covid-19 recovery. *Journal of Sustainable Finance & Investment*, 1–5.

Callegari, B., & Feder, C. (2022). A literature review of pandemics and development: The long-term perspective. *Economics of Disasters and Climate Change*, 1–30.

Cazcarro, I., Duarte, R., Sarasa, C., & Serrano, A. (2020). Perspectives on the Economics of the Environment in the Shadow of Coronavirus. *Environmental and Resource Economics, 76*(4), 447–517.

Cazcarro, I., García-Gusano, D., Iribarren, D., Linares, P., Romero, J. C., Arocena, P., Arto, I., Banacloche, S., Lechón, Y., Miguel, L. J., Zafrilla, J., López, L.-A., Langarita, R.,& Cadarso, M. Á. (2022). Energy-socio-economic-environmental modelling for the EU energy and post-COVID-19 transitions. *Science of the Total Environment*, 805. http://doi.org/10.1016/j.scitotenv.2021.150329

Chenet, H., Ryan-Collins, J., & van Lerven, F. (2021). Finance, climate-change and radical uncertainty: Towards a precautionary approach to financial policy. *Ecological Economics, 183*, 106957.

Chiappini, H. (2017). An introduction to social impact investing. In *Social impact funds* (pp. 7–50). Palgrave Macmillan.

Clemente-Suárez, V. J., Navarro-Jiménez, E., Moreno-Luna, L., Saavedra-Serrano, M. C., Jimenez, M., Simón, J. A., & Tornero-Aguilera, J. F. (2021). The impact of the COVID-19 pandemic on social, health, and economy. *Sustainability, 13*(11), 6314.

Climate Risk Is Investment Risk. (2020). *A fundamental reshaping of finance.*

Corfee-Morlot, J., Depledge, J., & Winkler, H. (2021). COVID-19 recovery and climate policy. *Climate Policy, 21*(10), 1249–1256.

Crowe, R. (2021). The EU recovery plan: New dynamics in the financing of the EU budget. In *The future of legal Europe: Will we trust in it?* (pp. 117–139). Springer.

Cumming, T., Seidl, A., Emerton, L., Spenceley, A., Kroner, R. G., Uwineza, Y., & van Zyl, H. (2021). Building sustainable finance for resilient protected and conserved areas: Lessons from COVID-19. *Parks, 27*, 149–160.

D'Orazio, P. (2021). Towards a post-pandemic policy framework to manage climate-related financial risks and resilience. *Climate Policy, 21*(10), 1368–1382.

Davies, R., Haldane, A. G., Nielsen, M., & Pezzini, S. (2014). Measuring the costs of short-termism. *Journal of Financial Stability, 12*, 16–25.

De Amorim, W. S., Valduga, I. B., Ribeiro, J. M. P., Williamson, V. G., Krauser, G. E., Magtoto, M. K., & de Andrade, J. B. S. O. (2018). The nexus between water, energy, and food in the context of the global risks: An analysis of the interactions between food, water, and energy security. *Environmental Impact Assessment Review, 72*, 1–11.

Deeg, R., & Hardie, I. (2016). What is patient capital and who supplies it? *Socio-Economic Review, 14*(4), 627–645.

de Vries, B. J. (2019). Inequality, SDG10 and the financial system. *Global Sustainability, 2*, e9, 1–2.

Echebarria Fernández, J. (2021). A critical analysis on the European Union's measures to overcome the economic impact of the COVID-19 pandemic. *European Papers—A Journal on Law and Integration, 5*(3), 1399–1423.

European Commission. (2020). *State of the Union Address by President von der Leyen.* https://ec.europa.eu/commission/presscorner/detail/en/SPEECH_20_1655

EU Technical Expert Group on Sustainable Finance. (2020). *5 high-level principles for Recovery & Resilience.* https://ec.europa.eu/info/sites/default/files/business_economy_euro/banking_and_finance/documents/200715-sustainable-finance-teg-statement-resilience-recovery_en.pdf

Faber, D. (2018). Global capitalism, reactionary neoliberalism, and the deepening of environmental injustices. *Capitalism Nature Socialism, 29*(2), 8–28.

Farzanegan, M. R., Feizi, M., & Gholipour, H. F. (2021). Globalization and the outbreak of COVID-19: An empirical analysis. *Journal of Risk and Financial Management, 14*(3), 105.

Fears, R., Gillett, W., Haines, A., Norton, M., & Ter Meulen, V. (2020). Post-pandemic recovery: Use of scientific advice to achieve social equity, planetary health, and economic benefits. *The Lancet Planetary Health, 4*(9), e383–e384.

Fernandes, C. I., Veiga, P. M., Ferreira, J. J., & Hughes, M. (2021). Green growth versus economic growth: Do sustainable technology transfer and innovations lead to an imperfect choice? *Business Strategy and the Environment, 30*(4), 2021–2037.

Fried, J. M., & Wang, C. C. (2019). Short-termism and capital flows. *Review of Corporate Finance Studies, 8*(1), 207–233.

Fuentes, R., Galeotti, M., Lanza, A., & Manzano, B. (2020). COVID-19 and climate change: A tale of two global problems. *Sustainability, 12*(20), 8560.

Furman, J. (2021). The crisis opportunity: What it will take to build back a better economy. *Foreign Affairs, 100*, 25.

Gary, S. N. (2019). Best interests in the long term: Fiduciary duties and ESG integration. *University of Colorado Law Review, 90*, 731.

Gillingham, K., & Sweeney, J. (2010). *Market failure and the structure of externalities* (pp. 69–91). RFF Press.

Goodell, J. W. (2020). COVID-19 and finance: Agendas for future research. *Finance Research Letters, 35*, 101512.

Gortsos, C. V. (2020). The response of the European central bank to the current pandemic crisis: Monetary policy and prudential banking supervision decisions. *European Company and Financial Law Review, 17*(3–4), 231–256.

Guerriero, C., Haines, A., & Pagano, M. (2020). Health and sustainability in post-pandemic economic policies. *Nature Sustainability, 3*(7), 494–496.

Gusheva, E., & de Gooyert, V. (2021). Can we have our cake and eat it? A review of the debate on green recovery from the COVID-19 crisis. *Sustainability, 13*(2), 874.

Henry, M. S., Bazilian, M. D., & Markuson, C. (2020). Just transitions: Histories and futures in a post-COVID world. *Energy Research & Social Science, 68*, 101668.

Hinarejos, A. (2020). Next generation EU: On the agreement of a COVID-19 recovery package. *European Law Review, 4*, 451–452.

Holling, C. S. (1986). The resilience of terrestrial ecosystems: Local surprise and global change. In W. C. Clark & R. E. Munn (Eds.), *Sustainable development of the biosphere* (pp. 292–317). Cambridge University Press.

Hughes, T. P., Barnes, M. L., Bellwood, D. R., Cinner, J. E., Cumming, G. S., Jackson, J. B., Kleypas, J., van de Leemput, I. A., Lough, J. M., Morrison,

T. H., Palumbi, S. R., van Nes, E. H., & Scheffer, M. (2017). Coral reefs in the Anthropocene. *Nature, 546*(7656), 82–90.

Hynes, W., Trump, B. D., Love, P., Kirman, A., Galaitsi, S. E., Ramos, G., & Linkov, I. (2020). Resilient financial systems can soften the next global financial crisis. *Challenge, 63*(6), 311–318.

International Monetary Fund. (2021). *Fiscal Monitor Update*. https://www.imf.org/en/Publications/FM/Issues/2021/03/29/fiscal-monitor-april-2021

Jamaludin, S., Azmir, N. A., Ayob, A. F. M., & Zainal, N. (2020). COVID-19 exit strategy: Transitioning towards a new normal. *Annals of Medicine and Surgery, 59*, 165–170.

Jena, P. R., Majhi, R., Kalli, R., Managi, S., & Majhi, B. (2021). Impact of COVID-19 on GDP of major economies: Application of the artificial neural network forecaster. *Economic Analysis and Policy, 69*, 324–339.

Johnston, D. W., Kung, C. S., & Shields, M. A. (2021). Who is resilient in a time of crisis? The importance of financial and non-financial resources. *Health Economics, 30*(12), 3051–3073.

Kalfagianni, A., & Papyrakis, E. (2022). Covid-19 and climate change. In *COVID-19 and international development* (pp. 147–156). Springer.

Kedward, K., Ryan-Collins, J., & Chenet, H. (2020). *Managing nature-related financial risks: A precautionary policy approach for central banks and financial supervisors*. Available at SSRN 3726637.

Kelly, M. P. (2021). The relation between the social and the biological and COVID-19. *Public Health, 196*, 18–23.

Khurshid, A., & Deng, X. (2021). Innovation for carbon mitigation: A hoax or road toward green growth? Evidence from newly industrialized economies. *Environmental Science and Pollution Research, 28*(6), 6392–6404.

Kołodko, G. W. (2020). After the calamity: Economics and politics in the post-pandemic world. *Polish Sociological Review, 210*(2), 137–156.

Kumar, A., & Ayedee, N. (2021). An interconnection between COVID-19 and climate change problem. *Journal of Statistics and Management Systems, 24*(2), 281–300.

Lagoarde-Segot, T. (2019). Sustainable finance: A critical realist perspective. *Research in International Business and Finance, 47*, 1–9.

Lagoarde-Segot, T., & Martínez, E. A. (2021). Ecological finance theory: New foundations. *International Review of Financial Analysis, 75*, 101741.

Lagoarde-Segot, T., & Paranque, B. (2018). Finance and sustainability: From ideology to utopia. *International Review of Financial Analysis, 55*, 80–92.

Lagoarde-Segot, T., Pérez, R., & Sun, W. (2021). Conclusion—finance and sustainability: An integrated thinking. In *Rethinking finance in the face of new challenges*. Emerald Publishing Limited.

Lahcen, B., Brusselaers, J., Vrancken, K., Dams, Y., Da Silva Paes, C., Eyckmans, J., & Rousseau, S. (2020). Green recovery policies for the COVID-19 crisis: Modelling the Impact on the economy and greenhouse gas emissions. *Environmental and Resource Economics, 76*(4), 731–750.

Langley, P. (2020). Impact investors: The ethical financialization of development, society and nature. In *The Routledge handbook of financial geography* (pp. 328–351). Routledge.

Laranja, M., & Pinto, H. (2022). Transformation for a post-pandemic world: Exploring social innovations in six domains. *Knowledge, 2*(1), 167–184.

Leach, M., MacGregor, H., Scoones, I., & Wilkinson, A. (2021). Post-pandemic transformations: How and why COVID-19 requires us to rethink development. *World Development, 138*, 105233.

Le Billon, P., Lujala, P., Singh, D., Culbert, V., & Kristoffersen, B. (2021). Fossil fuels, climate change, and the COVID-19 crisis: Pathways for a just and green post-pandemic recovery. *Climate Policy, 21*(10), 1347–1356.

Libecap, G. D. (2014). Addressing global environmental externalities: Transaction costs considerations. *Journal of Economic Literature, 52*(2), 424–479.

Luisetti, F. (2019). Geopower: On the states of nature of late capitalism. *European Journal of Social Theory, 22*(3), 342–363.

Loorbach, D., Schoenmaker, D., & Schramade, W. (2020). *Finance in transition: Principles for a positive finance future*. Rotterdam School of Management, Erasmus University.

Lydenberg, S. (2014). Reason, rationality, and fiduciary duty. *Journal of Business Ethics, 119*(3), 365–380.

Maida, L. (2021). Sustainable Finance—Integrating Sustainability into Corporate Banking. In *Corporate Sustainability in Practice* (pp. 111–124). Springer.

Mählmann, T. (2022). Negative externalities of mutual fund instability: Evidence from leveraged loan funds. *Journal of Banking & Finance, 134*, 106328.

Marginson, D., & McAulay, L. (2008). Exploring the debate on short-termism: A theoretical and empirical analysis. *Strategic Management Journal, 29*(3), 273–292.

Marlow, D., Pearson, L., MacDonald, D. H., Whitten, S., & Burn, S. (2011). A framework for considering externalities in urban water asset management. *Water Science and Technology, 64*(11), 2199–2206.

Marques, L. (2020). *Capitalism and environmental collapse*. Springer International Publishing.

Meltzer, J. P. (2016, September 21). Financing low carbon, climate resilient infrastructure: The role of climate finance and green financial systems. *Climate Resilient Infrastructure: The Role of Climate Finance and Green Financial Systems*.

Mishra, J., Mishra, P., & Arora, N. K. (2021). Linkages between environmental issues and zoonotic diseases: With reference to COVID-19 pandemic. *Environmental Sustainability, 4*(3), 455–467.

Mogi, R., & Spijker, J. (2021). The influence of social and economic ties to the spread of COVID-19 in Europe. *Journal of Population Research*, 1–17.

Monasterolo, I. (2020). Climate change and the financial system. *Annual Review of Resource Economics, 12*, 299–320.

Moreno, C., Allam, Z., Chabaud, D., Gall, C., & Pratlong, F. (2021). Introducing the "15-Minute City": Sustainability, resilience and place identity in future post-pandemic cities. *Smart Cities, 4*(1), 93–111.

Mukanjari, S., & Sterner, T. (2020). Charting a "green path" for recovery from COVID-19. *Environmental and Resource Economics, 76*(4), 825–853.

Murshed, S. M. (2020). Capitalism and COVID-19: Crisis at the Crossroads. *Peace Economics, Peace Science and Public Policy, 26*(3), 20200026.

Nesbitt, J. (2009). The role of short-termism in financial market crises. *Australian Accounting Review, 19*(4), 314–318.

Newell, P. (2022). Finance for the common good: Re-thinking the relationship between finance, poverty and sustainability. In *Financial crises, poverty and environmental sustainability: Challenges in the context of the SDGs and Covid-19 recovery* (pp. 17–24). Springer.

Newell, R., & Dale, A. (2020). COVID-19 and climate change: An integrated perspective. *Cities & Health, 5*, 1–5.

O'Callaghan, B. J., & Murdock, E. (2021). *Are we building back better? Evidence from 2020 and pathways for inclusive green recovery spending*. United Nations Environment Program.

O'Callaghan, B., Yau, N., Murdock, E., Tritsch, D., Janz, A., Blackwood, A., Purroy Sanchez, L., Sadler, A., Wen, E., Kope, H., Flodell, H., Tillman-Morris, L., Ostrovsky, N., Kitsberg, A., Lee, T., Hristov, D., Didarali, Z., Chowdhry, K., Karlubik, M., … Heeney, L. (2022). *Global Recovery Observatory*. Oxford University Economic Recovery Project. Accessed on 1 February 2022.

OECD. (2020). *OECD Business and Finance Outlook 2020 Sustainable and Resilient Finance*. OECD Publishing.

OECD. (2021). *The inequalities-environment nexus: Towards a people-centred green transition* (OECD Green Growth Papers, No. 2021/01). OECD Publishing.

OECD/IEA. (2021). *Update on recent progress in reform of inefficient fossil-fuel subsidies that encourage wasteful consumption 2021*. https://www.oecd.org/g20/topics/climate-sustainability-and-energy/OECD-IEA-G20-Fossil-Fuel-Subsidies-Reform-Update-2021.pdf .

Paranque, B., & Pérez, R. (2016). *Finance reconsidered: New perspectives for a responsible and sustainable finance*. Emerald Group Publishing Limited.

Penna, C. C., Schot, J., & Steinmueller, W. E. (2021). *The promise of transformative investment: Mapping the field of sustainability investing* (Deep Transitions Working Paper Series).

Pérez, R. (2021). Introduction: Finance, markets and society—Rethinking the paradigm. In *Rethinking finance in the face of new challenges*. Emerald Publishing Limited.

Perveen, N., Muzaffar, S. B., & Al-Deeb, M. A. (2021). Exploring human-animal host interactions and emergence of COVID-19: Evolutionary and ecological dynamics. *Saudi Journal of Biological Sciences, 28*(2), 1417–1425.

Pettifor, A. (2020). *The case for the green new deal*. Verso Books.

Pianta, S., Brutschin, E., van Ruijven, B., & Bosetti, V. (2021). Faster or slower decarbonization? Policymaker and stakeholder expectations on the effect of the COVID-19 pandemic on the global energy transition. *Energy Research & Social Science, 76*, 102025.

Piemonte, C., Cattaneo, O., Morris, R., Pincet, A., & Poensgen, K. (2019). *Transition Finance: Introducing a new concept* (No. 54). OECD Publishing.

Pinner, D., Rogers, M., & Samandari, H. (2020, April). Addressing climate change in a post-pandemic world. *McKinsey Quarterly*.

Ranger, N., Mahul, O., & Monasterolo, I. (2021). Managing the financial risks of climate change and pandemics: What we know (and don't know). *One Earth, 4*(10), 1375–1385.

Ranjbari, M., Esfandabadi, Z. S., Zanetti, M. C., Scagnelli, S. D., Siebers, P. O., Aghbashlo, M., Peng, W., Quatratro, F., & Tabatabaei, M. (2021). Three pillars of sustainability in the wake of COVID-19: A systematic review and future research agenda for sustainable development. *Journal of Cleaner Production, 297*, 126660.

Reyers, B., Folke, C., Moore, M. L., Biggs, R., & Galaz, V. (2018). Social-ecological systems insights for navigating the dynamics of the Anthropocene. *Annual Review of Environment and Resources, 43*, 267–289.

Rudebusch, G. D. (2021). Climate change is a source of financial risk. *FRBSF Economic Letter, 2021*(3), 1–6.

Ryszawska, B. (2018). Sustainable finance: Paradigm shift. In *Finance and Sustainability* (pp. 219–231). Springer.

Sanderson, H., Irato, D. M., Cerezo, N. P., Duel, H., Faria, P., & Torres, E. F. (2019). How do climate risks affect corporations and how could they address these risks? *SN Applied Sciences, 1*(12), 1–6.

San-Jose, L., Retolaza, J. L., & Van Liedekerke, L. (Eds.). (2021). *Handbook on Ethics in Finance*. Springer.

Schäfer, H. (2019). *On values in finance and ethics: Forgotten trails and promising pathways*. Springer.

Schumpeter, J. (1943). *Capitalism, socialism, and democracy*. Routledge.

Schwab, K. (2019). Why we need the "Davos Manifesto" for a better kind of capitalism. In *World Economic Forum* (Vol. 1).

Shiller, R. J. (2004). Radical financial innovation. *Entrepreneurship, Innovation, and the Growth Mechanism of the Free-Enterprise Economies, 306*, 316.

Shiller, R. J. (2013). *Finance and the good society*. Princeton University Press.
Sikora, A. (2021). European Green Deal—Legal and financial challenges of the climate change. In *Era Forum* (Vol. 21, No. 4, pp. 681–697). Springer Berlin Heidelberg.
Song, L., & Zhou, Y. (2020). The COVID-19 pandemic and its impact on the global economy: What does it take to turn crisis into opportunity? *China & World Economy, 28*(4), 1–25.
Steffen, B., Egli, F., Pahle, M., & Schmidt, T. S. (2020). Navigating the clean energy transition in the COVID-19 crisis. *Joule, 4*(6), 1137–1141.
Stensrud, A. B., & Eriksen, T. H. (Eds.). (2019). *Climate, capitalism and communities: An anthropology of environmental overheating*. Pluto Press.
Stevano, S., Franz, T., Dafermos, Y., & Van Waeyenberge, E. (2021). COVID-19 and crises of capitalism: Intensifying inequalities and global responses. *Canadian Journal of Development Studies/revue Canadienne D'études Du Développement, 42*(1–2), 1–17.
Stiglitz, J. E., Shiller, R. J., Gopinath, G., Reinhart, C. M., Posen, A., Prasad, E., Tooze, A., Tyson, L. D., & Mahbubani, K. (2020, April 15). How the economy will look after the coronavirus pandemic. *Foreign Policy*.
Storm, S. (2018). Financialization and economic development: a debate on the social efficiency of modern finance. *Development and Change, 49*(2), 302–329.
Sutton, R. T. (2019). Climate science needs to take risk assessment much more seriously. *Bulletin of the American Meteorological Society, 100*(9), 1637–1642.
Taherzadeh, O. (2021). Promise of a green economic recovery post-Covid: Trojan horse or turning point? *Global Sustainability, 4*, 1–19.
Taleb, N. N. (2012). *Antifragile: Things that gain from disorder*. Penguin.
Taleb, N. N., & Douady, R. (2011). *A map and simple heuristic to detect fragility, antifragility, and model error* (NYU-Poly Working Paper, SSRN).
Tian, J., Yu, L., Xue, R., Zhuang, S., & Shan, Y. (2022). Global low-carbon energy transition in the post-COVID-19 era. *Applied Energy, 307*, 118205.
Undorf, S., Tett, S. F., Hagg, J., Metzger, M. J., Wilson, C., Edmond, G., Jacques-Turner, M., Forrest, S., & Shoote, M. (2020). Understanding interdependent climate change risks using a serious game. *Bulletin of the American Meteorological Society, 101*(8), E1279–E1300.
United Nations Environmental Programme. (2011). *Towards a green economy: Pathways to sustainable development and poverty eradication*. UNEP/GRID-Arendal.
Vanhercke, B., & Verdun, A. (2022). The European semester as Goldilocks: Macroeconomic policy coordination and the recovery and resilience facility. *JCMS: Journal of Common Market Studies, 60*(1), 204–223.

Verdun, A., Vanhercke, B., & Spasova, S. (2022). Are (some) social players entering European recovery through the Semester back door? *Social Policy in the European Union: State of Play, 2021*, 107–130.

Verstein, A. (2017). Wrong-termism, right-termism, and the liability structure of investor time horizons. *Seattle University Law Review, 41*, 577.

Vitenu-Sackey, P. A., & Barfi, R. (2021). The impact of Covid-19 pandemic on the Global economy: Emphasis on poverty alleviation and economic growth. *The Economics and Finance Letters, 8*(1), 32–43.

Walsh, J., & O'Riordan, A. (2020). Financial markets and climate news: Evidence of short-termism? *Student Economic Review, 34*, 201–213.

Wei, X., & Han, L. (2021). The impact of COVID-19 pandemic on transmission of monetary policy to financial markets. *International Review of Financial Analysis, 74*, 101705.

Werikhe, A. (2022). Towards a green and sustainable recovery from COVID-19. *Current Research in Environmental Sustainability, 4*, 100124.

World Bank Group. (2019). *Boosting financial resilience to disaster shocks: Good practices and new frontiers*. World Bank Technical Contribution to the 2019 G20 Finance Ministers' and Central Bank Governors' Meeting, World Bank.

Wullweber, J. (2020, September 7). The COVID-19 financial crisis, global financial instabilities and transformations in the financial system. *Global Financial Instabilities and Transformations in the Financial System*.

Yanovski, B., Lessmann, K., & Tahri, I. (2020). *The link between short-termism and risk: Barriers to investment in long-term projects*. Available at SSRN 3550501.

Zabaniotou, A. (2020). A systemic approach to resilience and ecological sustainability during the COVID-19 pandemic: Human, societal, and ecological health as a system-wide emergent property in the Anthropocene. *Global Transitions, 2*, 116–126.

Zenghelis, D. (2021). Why sustainable, inclusive, and resilient investment makes for efficacious post-COVID medicine. *Wiley Interdisciplinary Reviews: Climate Change, 12*(4), e708.

Ziolo, M., Filipiak, B. Z., Bąk, I., Cheba, K., Tîrca, D. M., & Novo-Corti, I. (2019). Finance, sustainability and negative externalities: An overview of the European context. *Sustainability, 11*(15), 4249.

CHAPTER 7

Conclusions

7.1 Introduction

The green financing framework enables investors to support delivery organizations in developing and developed countries, in order to address environmental challenges by facilitating investing in many asset classes. In this process, they receive adequate, or in some cases, below-market financial returns, through a wide range of innovative financial tools. A number of advances in policy, regulations, and market innovations are transforming the sector, especially in the European Union context. In this field, the clarification of what constitutes green investing and what "green" means in this context is crucial for the definitive advancement of the sector.

Regarding green finance delivery tools, emerging markets are experimenting with financial innovations that undoubtably encourage investments in green and low-carbon activities; this helps to reduce the financial gap in aligning actions with the trajectory of the Paris Agreement, and provides capital for economic activities that assist the transition. However, several issues affect the ability of such instruments and marketplaces to realize their potential to accelerate the decarbonization of the real economy at the required speed.

Also for non-targeted environmental investments, emerging frameworks such as (green) SDG portfolio alignment should help these strategies to overcome a series of challenges, such as a lack of standardization and disclosure frameworks; and to capture the *patience-averse* (but environmentally oriented) mainstream investors. Furthermore, *greenwashing* is a prevalent issue, affecting both green financing instruments and asset managers. Its challenges require a set of measures that address different targets, and the adoption of various strategies for regulatory actions.

In this evolving context, the nexus between finance and sustainable environmental development and protection requires new ways of thinking, both theoretically and practically. Especially after the COVID-19 pandemic, finance is gaining a new role in this transitional process. A synthesis of theoretical contribution delivered in this book, along all the previous chapters is illustrated in Fig. 7.1.

However, many questions remain unanswered, especially in the academic field. This chapter, therefore, outlines directions for future research and provides suggestions to foster growth in the green finance industry.

7.2 The Definitional Puzzle: Identifying Green Beyond the "What"

The definition of green finance has evolved over time, shifting from macro considerations related to a vision of finance that promotes environmentalism, to the greening of the financial system to tackle environmental issues. After the Paris Agreement in 2015, research in green finance has entered a vigorous stage, and the focus of green finance definitions has started to integrate environmental factors into investment decisions, for long-term value creation. However, a certain level of ambiguity remains, due to the existence of overlapping terms. Indeed, "sustainable finance," "green finance," and "climate finance" are often used interchangeably, and are therefore misunderstood. Moreover, various stakeholders to identify green finance sometimes use other terms such as "ESG investments and responsible investments."

Furthermore, the definitional puzzle of green finance is related to a second dimension: the identification of what "green" signifies in the context of green finance. Resolving this issue is crucial for the advancement of the field. Indeed, all the frameworks, taxonomies, and eligibility criteria adopted for green ratings and the use of proceeds are based on different conceptions of green activities. Within this arena, the emergence of transition finance represents a further element of complexity, which is reflected in difficulties in regulatory actions, standardization efforts, and

7 CONCLUSIONS 161

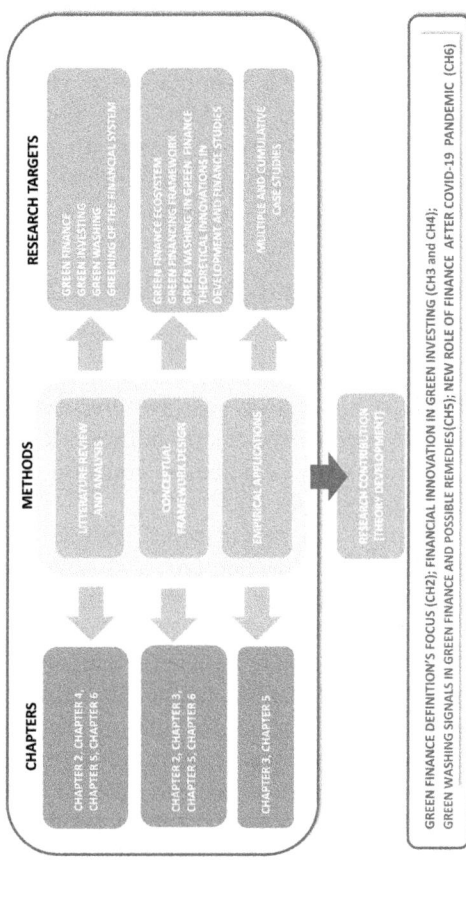

Fig. 7.1 A sysntesis of the book's research contribution (*Source* Author's elaboration)

finally, in the correct market sizing of the green finance sector. A wider agreed definition of green finance is therefore central to the development of green finance as an academic research field, as well as a market segment within the broader sector of finance for sustainable development.

As a suggestion for future research in this field, empirical studies should be prioritized, to provide comparisons and analysis derived from the observation of market practices and regulatory frameworks; this will help policymakers to define standardized, widely agreed standards. However, this transition needs to overcome the dominance of the taxonomical method (based on "what" investments address) in the identification of green activities, by designing frameworks based on the integration of impact-based metrics (which identify investment *for* a precise and measurable environmental benefit).

7.3 The Future of Green Investing Beyond Taxonomies and Ratings

The green investment sector has evolved in recent years, moving from simple exclusion, to ESG's integration within investment decision-making. More recently, the green investment framework has used the ESG lens to focus on companies whose activities contribute to achieving the green Sustainable Development Goals.

The ESG sector is complex and multifaceted. Actors and portfolio strategies rely on different ESG data outcomes, mainly derived from companies' sustainability reporting. Thus, the harmonization of sustainability reporting frameworks and standards will be crucial for the advancement of this sector, through the simplification of sustainability reporting and reducing ambiguity in the related ESG analysis. In this sense, the EU market represents a promising field for improvement in the ESG market, due to its recent regulation innovations, mainly in terms of corporate sustainability disclosure. On the other hand, the (green) SDG investing trend has added new considerations to the existing prominent investment approaches, such as exclusionary screening and ESG integration; moreover, it has added a new investment segment that is "impact-aligned" to the SDGs. In simpler terms, SDGs provide a common language for including environmental impact issues in the value creation chain of green investments. Therefore, an investment approach that focuses on the achievement of SDG outcomes requires an analysis that moves beyond the financial materiality of ESG issues at the individual investee level, to adequately capture the outcomes for society and the environment at a systems level.

This emerging trend has interesting implications for the advancement of the green finance market, for various reasons. First, the integration of environmental sustainability and impact via the SDGs is just one trend shaping the climate–environment conversation within funds. Furthermore, this approach provides an additional layer of transparency, for investors to better assess the *greenness* of their portfolio companies. In this light, there are abundant future research avenues for addressing these evolving approaches; especially by providing empirical insights and theoretical frameworks that can be applied in the equity green finance segment.

7.4 Green Finance After Covid: The New Normal for the Financial System?

The increasing climate of uncertainty, deriving from repeated economic and financial shocks, began in the twenty-first century with the Twin Towers attack, followed by the 2008 financial crisis. The recent wave produced by the COVID-19 pandemic has brought a growing consensus that the socio-economic system is not a separate entity from the financial system. Consequently, a consistent body of research has recognized the need to overcome neoclassical financial theory, by theorizing that a radical change in the financial system is required to address this series of shocks. As a result, scholars have increasingly focused on the necessity for innovation in financial theory, by acknowledging the role of values, ethics, and ideologies in finance studies.

These urgent and disruptive theoretical innovations have propitiously coincided with a new era of "re-thinking" theoretical economic development assumptions; this has drawn attention to the economic and financial system's ability to grow sustainably, while also increasing its *antifragility* in the face of repeated and severe shocks. This scenario should include transition finance, meaning the adoption of a holistic approach to finance for green growth, rather than being reduced to a *finance for green virtue*, which could dangerously diverge from a systemic change by completely isolating carbon industries in their transformation journey activities. Finally, at both a macro and a micro level, the definitive affirmation of integrating long-term environmental values into investment decisions will have decisive results.

In this evolving situation, regulatory authorities, industry associations, banks, and fund managers should work closely in designing a new role for finance, which takes the following into consideration: (*i*) the inclusion, within the systemic risks, of climate and environmental risks; (*ii*) the inclusion of finance for *transition* within green-targeted financial investments; and (*iii*) the definitive integration of environmental factors into modern frameworks of investors' fiduciary duties. Future studies in this field should be devoted to understanding such emerging practices, and establishing the scientific maturity of these converging trends.

7.5 Conclusion

This book has discussed many aspects of green finance, by addressing some existing questions and raising further ones. It has clearly revealed the need for an academic response to the tumultuous transformation facing traditional finance, and future explorations of some emerging topics will be useful in this light. In particular, two main areas of research are emerging: the shifting focus toward the consideration of environmental impact in investment decisions, and solutions to the problem of greenwashing. These challenges will also require the contribution of all market players in the green financial industry.

Index

A
Adaptation finance, 17
Anti-fragile, 142
Asset classes, 5, 11, 42, 68, 116, 159
Asset management, 24
Asset manager, 23, 24, 58, 93, 113, 114, 119, 121, 122, 124, 147, 148, 160

B
Banking, 24, 57, 63, 117, 121, 136
Banks, 12, 15, 23, 24, 26, 27, 29–31, 38, 40, 41, 56, 57, 60, 62, 64, 93, 111, 117, 121, 136, 164
Best-in-class, 30, 40, 58, 91
Biodiversity, 17, 21, 39, 40, 61, 111, 139
Blue finance, 17
Business model, 1, 14, 21, 23, 24, 55, 57, 65, 72, 145, 146

C
Capital Market Union (CMU), 13
Carbon finance, 20, 30, 63
Circular economy, 21, 41, 61, 111, 115, 135
Clean energy, 18, 29, 39, 86, 88
Climate Bonds, 29
Climate finance, 17, 20, 31, 40, 42, 63, 64, 160
Conservation finance, 17, 21
COP21, 9, 12
Corporate Social Responsibility (CSR), 23, 107, 109, 110
COVID-19, 1, 2, 10, 14, 55, 93, 133, 135–138, 141, 145, 147, 148, 163
Crowdfunding, 25, 65, 71, 72, 76

D
Debt, 11, 25, 26, 29, 30, 38, 56, 59, 62, 71, 72, 90, 92
Development Finance Institution (DFIs), 24, 62

INDEX

E

Environmental finance, 16, 17
Environmental impact, 10, 14, 25, 27, 28, 30, 35, 40, 58, 68, 69, 72, 74, 76, 88, 92, 100, 124, 127, 128, 138, 143, 147, 162, 164
Environmental investing, 11, 16, 17
Environmental, Social and Governance (ESG), 12, 16, 18, 38–40, 42, 58, 59, 69, 85, 86, 91–98, 100, 112–114, 116, 119, 120, 124, 126, 147, 162
Environmental sustainability, 2, 5, 27, 34, 59, 61, 62, 87, 100, 117, 163
Equity, 25, 26, 29, 30, 39, 40, 56, 59, 62, 63, 65, 66, 72, 90, 91, 98–100, 126, 128, 163
ESG Index, 95
ESG investing, 18, 86, 92, 93, 128
ESG rating, 58, 93–96
EU Green Deal, 1, 10, 35, 139
EU non-financial reporting directive, 1
European Union (EU), 10, 14, 18, 24, 32, 35, 36, 96, 108, 111, 112, 115–117, 127, 134, 137–139, 148, 159, 162
Exchange Traded Funds (ETFs), 58, 99, 100

F

Fiduciary duty, 34, 147
Financial disclosure, 116
Financial industry, 1, 5, 6, 10, 24, 33, 108, 164
Financial intermediaries, 24, 56
Financial sector, 11, 31, 56, 85, 92, 93, 108, 110, 111, 121, 127, 136, 142, 144, 149
Financial system, 3, 5, 13–15, 17, 24, 31, 34, 42, 113, 134, 136, 137, 140–143, 145, 148, 160, 163
Financial theory, 5, 10, 140, 143, 144, 163
Forest finance, 17, 21
Funds, 23–26, 28–31, 39, 40, 57, 58, 62–64, 66, 72, 73, 91, 92, 98, 100, 112–114, 116, 119, 121, 124–126, 134, 137–139, 144, 163, 164

G

Green bank, 31
Green banking, 57
Green bond, 23–25, 27–29, 36–38, 57–62, 69, 70, 75, 90, 92, 112, 116, 117, 122, 124, 139
Green bond principles (GBP), 27, 28, 33, 36, 37, 60
Green business, 1, 19, 23, 55
Green economy, 10, 13, 55, 145
Green finance (GF), 1–6, 9–19, 21, 24–26, 29–31, 34–36, 39, 41, 42, 55, 56, 62–65, 71, 76, 92, 95, 98, 99, 108, 114–116, 119, 121–124, 126–128, 141, 143, 144, 146, 147, 159, 160, 162–164
Green finance ecosystem, 22, 24, 25, 121
Green finance market, 5, 11, 31, 35, 36, 39, 56, 57, 59, 100, 120, 128, 144, 163
Green finance standards, 31, 34
Green financing framework, 3, 56, 59, 159
Green investing, 4, 63, 85, 86, 99, 147, 159
Green loans, 26, 27, 30, 38, 64, 92, 116, 127
Green mortgages, 27

Green portfolio, 5, 124
Green recovery, 135, 138
Greenwashing, 2–6, 62, 70, 94, 107–110, 112–114, 117, 119, 121–128, 160, 164

H
High Net Worth Individuals (HNWI), 23, 29

I
Impact, 2, 18–21, 30, 40, 58, 62, 66, 68, 69, 72–74, 85, 90, 91, 94, 96–100, 109, 110, 113, 122, 124, 126, 133, 135–137, 144, 145, 147, 148, 163
Impact bond (IB), 68, 71, 73, 74, 76
Impact investing (II), 30, 40, 66, 68, 92, 98, 127, 143, 147
Institutional investor, 23, 56, 57, 93, 120

K
Key performance indicators (KPIs), 69, 98, 100, 124

L
Low carbon, 124

M
Mitigation finance, 17

N
Negative screening, 40
Net-zero emissions, 115, 146

P
Paris Agreement, 12, 28, 42, 62, 76, 115, 134, 146, 159, 160
Portfolio alignment, 159
Post-COVID, 134, 137

R
Renewable energy, 16, 19, 20, 24, 27, 29, 39, 61, 62, 65, 90, 98, 135
Responsible investment, 13, 42, 120, 160
Risk, 2, 3, 5, 12, 15, 25, 26, 34, 58, 59, 61, 68, 72–74, 88, 90–95, 97–99, 107, 108, 112–117, 119–121, 126–128, 136, 139, 144, 145, 147, 148, 164

S
Shareholders, 24, 29, 57, 110
Short termism, 148
Stakeholder, 24, 31, 42, 57, 74, 86, 107–110, 115, 120, 121, 124, 160
Sustainability-linked bonds (SLBs), 36, 64, 69, 71, 75, 76, 124
Sustainable development, 10–13, 15, 42, 58, 85, 87, 88, 134, 162
Sustainable Development Goals (SDGs), 1, 3, 5, 9, 12, 21, 86, 87, 92, 96–100, 146, 159, 162, 163
Sustainable finance, 6, 13, 14, 16, 18, 19, 35, 42, 71, 112, 160
Sustainable investment, 10, 91, 93, 111–113, 124
Systemic risk, 2, 142, 144, 145, 164

T
Taxonomy, 10, 14, 28, 31, 35, 42, 60, 76, 96, 108, 110–112,

114–118, 123–125, 128, 139, 143, 146, 160

Thematic investing, 98

Transition, 3, 10, 13–15, 17, 18, 21, 31, 36, 55, 73, 76, 87, 111, 112, 114, 115, 121, 124, 134, 136, 137, 145, 146, 159, 162

Transition finance, 145, 146, 160, 163

2030 agenda, 86

U

United Nations Environment Programme (UNEP), 11, 40

United Nations (UN), 1, 86, 146

Use of proceeds, 2, 26–28, 37, 38, 42, 60–62, 69, 90, 127, 146, 160

GPSR Compliance

The European Union's (EU) General Product Safety Regulation (GPSR) is a set of rules that requires consumer products to be safe and our obligations to ensure this.

If you have any concerns about our products, you can contact us on

ProductSafety@springernature.com

In case Publisher is established outside the EU, the EU authorized representative is:

Springer Nature Customer Service Center GmbH
Europaplatz 3
69115 Heidelberg, Germany

www.ingramcontent.com/pod-product-compliance
Ingram Content Group UK Ltd.
Pitfield, Milton Keynes, MK11 3LW, UK
UKHW021251180426

11946UKWH00004B/79